Optimizing
The Soil Physical Environment
Toward Greater Crop Yields

*Proceedings of an Invitational Panel
convened during the International
Symposium on Soil-Water Physics and Technology
held at The Hebrew University Faculty of
Agriculture in Rehovot, Israel,
August 29 to September 5, 1971*

Optimizing
The Soil Physical Environment
Toward Greater Crop Yields

EDITED BY

DANIEL HILLEL

Department of Soil Science
The Hebrew University of Jerusalem
Rehovot, Israel

Academic Press   **New York and London**   **1972**

ACADEMIC PRESS, INC.
111 Fifth Avenue, New York, New York 10003

United Kingdom Edition published by
ACADEMIC PRESS, INC. (LONDON) LTD.
24/28 Oval Road, London NW1 7DD

LIBRARY OF CONGRESS CATALOG CARD NUMBER: 71-182633

PRINTED IN THE UNITED STATES OF AMERICA

CONTENTS

CONTRIBUTORS ix
PREFACE xi
ACKNOWLEDGMENTS xiii
PROLOGUE: Soil Physics and Technology: The Tasks before Us xv
    D. Hillel

**Criteria for Determining the Aims and Directions of Research
in Soil Physics and Technology**
    W. R. Gardner
    General 1
    References 9

Efficient Management of Water in Agriculture
    T. J. Marshall
    Irrigation Development 11
    Irrigation Efficiency 12
    Drainage 14
    Salinity of Soil and Water 15
    Control of Soil Water 18
    Management in Relation to the Environment 20
    References 21

Soil Temperature and Crop Growth
    C. H. M. van Bavel
    Introduction 23
    Current State of Theoretical Concepts 24
    Agronomic Significance of Soil Temperature Regimes 25
    Relating Crop Response to the Soil Temperature Regime 28
    Conclusions 31
    References 32

Improving The Water Properties of Sand Soil
    A. E. Erickson
    General 35
    References 40

Improvement of Soil Structure by Chemical Means
M. De Boodt
Introduction 43
Soil Physical Growth Factors 44
Modification of Soil Physical Conditions 45
Solutions of Polymers as Soil Conditioners 48
Emulsions of Polymers as Soil Conditioners 50
Application of Soil Conditioning Emulsions in the Field 52
Summary and Conclusions 53
References 54

Root Development in Relation to Soil Physical Conditions
H. M. Taylor, M. G. Huck, and B. Klepper
Introduction 57
Root Growth 57
Soil Temperature 59
Soil Aeration 61
Soil Water Status 65
Soil Strength 67
Interaction of Root Growth Factors 73
Conclusions 74
References 75

The Field Water Balance and Water Use Efficiency
D. Hillel
The Field Water Balance 79
Evaluation of the Water Balance 81
Water Use Efficiency 90
Possibilities of Control 91
Need for Overall Control of the Plant Environment 96
References 97

Control of Soil Salinity
E. Bresler
Effect of Soil Salinity on Plants 102
Solute Movement in Soils 104
Effect of Soil Salinity on Water Flow 109
Estimation of Salt Distribution in the Soil Profile under
    Fallow Conditions 112
Estimation of Salt Distribution in the Soil Profile under
    Crop Growing Conditions 115
Short-Run Soil Salinity Control 119

CONTENTS

Long-Run Salinity Control 124
Economic Approach to Short-Run Salinity Control 126
References 128

Programming Irrigation for Greater Efficiency
   M. E. Jensen

Introduction 133
Alternative Methods of Improving Irrigation Programming 135
Irrigation Management Service Requirements 136
Meteorological Data Limitations 139
Effective Use of Regional Experimental Data 141
USDA-ARS-SWC Irrigation-Scheduling Computer Program 141
Summary 158
References 158

Water Utilization by a Dryland Rowcrop
   H. R . Gardner

General 163
References 171

The Control of the Radiation Climate of Plant Communities
   M. Fuchs

Introduction 173
The Solar Radiation Balance in a Vegetation Stand 176
Control of the Diffuse Radiation Climate 180
Control of the Radiation Climate for Beam Incidence 184
Conclusions 189
References 190

Nutrient Supply and Uptake in Relation to Soil Physical Conditions
   R. E. Danielson

Introduction 193
Nutrient Supply 195
Available Forms 196
Nutrient Contact with Root Surfaces 197
Nutrient Uptake and Use by Plants 202
Managing the Soil Physical Condition for Improved
   Plant Nutrition 204
Conclusions 216
References 217

CONTENTS

Analysis of Growth Parameters and Their Fluctuations in Searching for Increased Yields*

D. Zaslavsky

The Medium We Work with 223
Possible Effects of Fluctuations on Plant Yields 225
Multi-Variate Fluctuations 227
Net Change in Equilibrium Conditions due to Fluctuations 229
Relaxation of Hysteresis 229
Increased Apparent Conductivity 230
Some Hydrological Conjectures 230
Conclusions 231
References 232

SUBJECT INDEX 233

CONTRIBUTORS

Bresler, E., Division of Soil Physics, Volcani Institute of Agricultural Research, Beit Dagan, Israel

Danielson, R. E., Department of Agronomy, Colorado State University, Fort Collins, Colorado

De Boodt, M., Leerstoel voor Bodemfysika, Fakulteit van de Landbouw-wetenschappen, Rijksuniversiteit, Gent, Belgium

Erickson, A. E., Department of Crop and Soil Sciences, Michigan State University, East Lansing, Michigan

Fuchs, M., Division of Agricultural Meteorology, Volcani Institute of Agricultural Research, Beit Dagan, Israel

Gardner, H. R., Northern Plains Branch, Soil and Water Conservation Research Division, Agricultural Research Service, U. S. Department of Agriculture, Fort Collins, Colorado

Gardner, W. R., Department of Soil Science, University of Wisconsin, Madison, Wisconsin

Hillel, D., Department of Soil Science, Faculty of Agriculture, The Hebrew University of Jerusalem, Rehovot, Israel

Huck, M. G., Soil and Water Conservation Research Division, Agricultural Research Service, U. S. Department of Agriculture, Auburn University Agricultural Experiment Station, Auburn, Alabama

Jensen, M. E., Snake River Conservation Research Center, Soil and Water Conservation Research Division, Agricultural Research Service, U. S. Department of Agriculture, Kimberly, Idaho

Klepper, B., Department of Botany and Microbiology, Auburn University Agricultural Experiment Station, Auburn, Alabama

Marshall, T. J., Division of Soils, Commonwealth Scientific and Industrial Research Organization, Adelaide, Australia

Taylor, H. M., Soil and Water Conservation Research Division, Agricultural Research Service, U. S. Department of Agriculture, Auburn University Agricultural Experiment Station, Auburn, Alabama

van Bavel, C. H. M., Department of Soil and Crop Sciences, Texas A & M University, College Station, Texas

Zaslavsky, D., Soil and Water Faculty of Agricultural Engineering, Technion–Israel Institute of Technology, Haifa, Israel

PREFACE

This book presents papers that were given at a special invitational panel on the topic "Optimizing the Soil Physical Environment toward Greater Crop Yields," convened as part of the International Symposium on Soil-Water Physics and Technology which was held at The Hebrew University Faculty of Agriculture in Rehovot, Israel, between August 29 and September 5, 1971. The Symposium was sponsored jointly by the International Soil Science Society (Soil Physics and Soil Technology Commissions) and the Israel Society of Soil Science.

The idea of the panel was to bring together some of the most active researchers in applied soil physics in an attempt to trigger a common effort to summarize and evaluate the current status and trends of research in this vital field. The growing need for more comprehensive and integrated information on the natural behavior and agricultural management of the soil-plant-atmosphere system is universally and keenly felt. Such is the complex and dynamic nature of this system that only by coordinating the contributions of many scientists with different, though complementary, points of view, can we hope to obtain the necessary integrated information.

Accordingly, the contributions presented at the panel and published herein cover a wide range of topics: from a critical and somewhat philosophical questioning of our current goals and criteria for selecting and initiating research projects; through basic elucidations of the transformations and fluxes of energy and matter in the field; to methods of measuring, managing, and modifying the crop production system to greater agricultural advantage. Whether these diverse contributions indeed intermesh to provide a consistent and comprehensive view of what is in nature a unified environment is for the readers to judge. Inevitably, such a book reflects (as it very well should) not only what is known, but also what is missing in our present-day incomplete conception of that environment.

D. Hillel

ACKNOWLEDGMENTS

Thanks are due to the following individuals and groups for their direct and indirect contributions to the Panel herein published and the preparation of the book itself:

The sponsors of the International Symposium on Soil-Water Physics and Technology: Professor F. A. van Baren, Secretary-General of the International Soil Science Society; Professor W. R. Gardner, President, Commission I (Soil Physics), International Soil Science Society; Dr. T. J. Marshall, President, Commission VI (Soil Technology), International Soil Science Society; and Dr. A. Banin, President, Soil Science Society of Israel.

The Organizing Committee of the International Symposium on Soil-Water Physics and Technology, under the chairmanship of Dr. J. Shalhevet (Volcani Institute of Agricultural Research, Beit Dagan, Israel) and the Kenes Organization for the great effort of helping to arrange the Symposium within which the Panel herein published took place.

The Faculty of Agriculture of The Hebrew University of Jerusalem, and particularly Dean S. Monselise and Associate Dean N. Shaffer, for hosting the Symposium and the Panel.

Mr. S. Tal for his careful work in drawing many of the figures.

Last but not least, Naomi Revzin for her painstaking and meticulous work in typing the book. In so doing, she has had to contend with and overcome numerous difficult passages and editorial idiosyncracies. The appearance of this book is evidence of her monumental patience and perserverance.

PROLOGUE

SOIL PHYSICS AND TECHNOLOGY:
THE TASKS BEFORE US

It seems singularly appropriate that the panel published herein was held in Israel, this narrow strip of land locked between the jaws of sea and desert, where the issue of soil and water has always been crucial and inescapable. Here a man can walk from a humid region of wooded hills across the edge of life into the parched desert in little more than half an hour. In this very ancient land where agriculture likely originated, the struggle of the desert against the sown has been raging since time immemorial.

Here, indeed, the importance of the soil-water system in nature and in the life of man has been realized since the dawn of civilization and man's awakening awareness of his relationship to his environment.

In the Hebrew Bible, the account of creation describes how "the Spirit of God moved upon the face of the waters," and how the waters were divided and separated from dry land; then how the waters were made to bring forth an abundance of life, and how man himself was created out of, and is destined to return to, "affar," which is, literally, the material of the soil. The second chapter of Genesis describes two divine methods of irrigation. The first description tells of how "there was no rain upon the earth, but there went up a mist from the earth and watered the face of the ground" (a method of water conservation by recycling the vapor in a closed system, indeed a method we ought to emulate today!). The second description tells of how a river was made to water the Garden of Eden, and how it was divided into four heads.

In the Book of Deuteronomy, the Land of Promise is described as "a good land, a land of brooks of water, of fountains and depths that spring out of valleys and hills." And, further, that "I will give you the rain of your land in due season, the first rain and the latter rains, that thou mayest gather in thy corn and thy wine and thine oil." But the abundance of the land was never to be taken for granted, as the Book admonishes: "Take heed to yourselves, that your heart be not deceived, and ye turn aside and worship other gods" (perhaps the gods of unbridled commercial development and mindless exploitation and despoliation of the environment?). "Then the Lord's wrath be kindled against you, and he shut up the heaven that there be no rain, and that the land yield not her fruit."

The whole Bible, in fact, is replete with references to soil and water. In the Second Book of Samuel, the wise woman speaks of wasted lives in terms of "water spilt on the ground, which cannot be gathered up again."

The Book of Proverbs might even have anticipated this international conference as it stated: "As cold water to a thirsty soul, so is good news from a far country." And when wise old King Solomon said: "Stolen waters are sweet," just what did he mean?

Finally, we are left to ponder the terse summary of the whole hydrological cycle, as given in Ecclesiastes: "All the rivers run into the sea, yet the sea is not full; unto the place whence the rivers come, thither they return again."

H. G. Wells, the English historian and thinker, once said that the history of mankind is a race between learning and disaster. The form of disaster which threatens mankind changes like a many-headed dragon. Once it was pestilence which decimated entire regions. Then, as medicine learned to control the diseases one after another, it inadvertently set the stage for an unprecedented growth of population which created a host of new and unforeseen threats. The threat of hunger is still upon us, not yet completely alleviated by the development of agriculture. Already, however, this development has resulted quite unexpectedly in new threats of environmental damage.

By its very character, agriculture interferes with the environment. In nature, there is free competition among biology's many forms and species. The moment he carves out a tract of land and turns it into a crop field, man coerces nature to do his bidding by favoring one species over others. To encourage the growth of his crops, man suppresses all other species he considers undesirable. Yet in an open system such species continuously fight back and reinvade their stolen realm. To combat them, man has developed mechanical and later chemical means. Initially these resulted in tremendous increases in production efficiency. Such is the irony of human fate, however, that what was once hailed as a pure blessing has gradually begun to develop into the threat of a new disaster. The field, still open to and part of a larger environment, transmits its poisonous residues and products to the air, water, animals, or plants. Eventually, these poisons might turn against man himself.

To combat the new threat, a new level of knowledge is required, obtainable only by the occasionally inspired but often arduous and painstaking process of research.

The gathering danger of environment deterioration has further complicated the task of increasing agricultural production, which by itself is of growing urgency.

In many countries, and certainly in Israel, agriculture depends primarily on irrigation, and is by far the largest user of water. Let us therefore consider the problem of water. Here our concern derives from a growing worldwide shortage in the midst of seeming plenty. For although water submerges nearly three-fourths of the earth's surface, it is unfortunate that it cannot always be found where, when, and in the amount and quality needed. 97% of the earth's water is in seas and oceans, too salty for drinking or irrigation. Another 2% is locked in glacier and ice caps. The remaining fraction is neither evenly distributed nor properly used. In the next twenty years, the world's demand for water will double as cities grow and as industry and agriculture expand everywhere. It still takes twenty tons of water to refine one ton of petroleum, up to 250 tons of water to make one ton of steel, and 1,000 tons of water to produce one ton of grain!

One answer to the water shortage is desalination. It is, incidentally, one of the oldest methods. Aristotle taught that "salt water, when it turns into vapor, becomes sweet, and the vapor does not form salt water when it condenses." Julius Caesar used stills to convert sea water for his thirsty legions during the siege of Alexandria. However, only now is desalination being attempted on a large scale, and it will still be some years before its use becomes widespread. Even when it becomes economically feasible in terms of water cost, there will still be problems of conveying it inland in many locations. Desalination is not the panacea that the public has been led to expect very soon.

The supply and distribution of water may some day be controlled by weather modification. Cloud seeding is a first step in that direction. But at present we still cannot predict the weather, let alone manage it.

For the time being, we are faced with the inexorable reality that the supply of fresh water, though indispensable, is strictly limited. It is for this reason that water has inevitably caused disputes. The very word "rival" was used in Roman law as a term for those who shared the water of a *rivus*, or irrigation channel. Where we cannot increase the supply, we must at least avoid waste and increase the efficiency of water utilization.

Nearly half of the world's population is struggling to eke out a living in the world's semi-arid and arid zones, where, by a cruel stroke of nature, the requirements of most plants for water are greatest even while the supplies of water by rainfall are least. This discrepancy strongly affects agricultural plants, which by their physiology and interaction with the field environment must transpire constantly while they are productive, and in fact transmit hundreds of times more water to the dry air than they strictly need for growth. Thus in the arid zones the scales are weighted heavily against agriculture, and the balance must be rectified by intensive irrigation and water conservation.

It is the ability of the soil to serve as a reservoir for water which must bridge the gap between climatically induced transpiration, which is practically incessant, and the supply of water, which is intermittent and infrequent. But the soil is a leaky reservoir which loses water downward by seepage and upward by evaporation. To bring the plight of the plants closer to the understanding of the public, we might describe them as living under a regime which taxes not 30 or 40% of their income as our governments might tax us, but 99% or more of their water intake. Moreover, imagine the plants as having to depend on a bank that is robbed and embezzled almost daily... . The function of agriculture is to produce chemicals — sugar, starch, oil, protein, cellulose, etc. Now imagine what chemical plant could operate efficiently if its water supply could be shut off at any time without notice! The utilization of energy in the field is no more efficient than that of water, as more than 90% of the sun's radiant energy is generally dissipated in processes other than photosynthesis, with evaporation again being the prime user.

These are some of the basic problems confronting the relatively new science of soil physics and technology, which really only came into its own some thirty years ago. To the soil physicist, the field stretching outside his window is not the serene place of pastoral poets, but a dynamic system in a constant state of turmoil in which matter and energy are in unceasing flux. Some of the questions which this system poses and which soil physicists and technologists must answer involve the fate and disposition of the sun's energy as it reaches the field, the interactions it induces in the soil and water stored in it, and the effects of the water, solute, and energy fluxes on the productive processes of plants. To be useful, the knowledge required of this dynamic system must be quantitative. We must be able to measure and relate the variable rates of simultaneous processes, and to predict how these rates might change under possible control measures. The physical requirements of agricultural production have been known qualitatively for some time, but in quantitative terms the answers available to date are as yet only approximate, and sometimes grossly so. Even where we have exact knowledge, it sometimes pertains to an idealized rather than the real system. We have yet, for instance, to really apply our very exact theories of infiltration to the actual field or watershed, which is heterogeneous and three-dimensional.

That better knowledge is imperative to ensure the future of our environment and our agriculture has been proven long ago by past failures, in the ancient Middle East as well as in modern and highly developed countries. In the United States of America, for instance, it has been estimated that all new irrigation projects now under implementation will scarcely make up for the amount of land withdrawn from production owing to deterioration. In some arid and semi-arid countries, we can find frightening examples of once-thriving agricultural fields reduced to desolation by practically irreversible salinization caused by injudicious management of the soil-water system. We can ill afford to have this happen to us in Israel, since we have nowhere else to go.

Israel is not alone in this situation. The problems are truly universal, as is. attested by the international interest they engender and the efforts toward their solution in many countries. Yet we submit that these efforts are not enough. The potential and need for international collaboration are still unrealized.

We as soil scientists are among the fortunate few who have the opportunity to participate in a worldwide effort which will transcend the narrow and divisive boundaries of institutions and states. To rise to this task, we must make some special effort not to be lulled into complacency in our comfortable little niches. Our true vocation is a constant restless quest, search and re-search. Part of the quest is to be individual, but much of it must be pursued cooperatively. No one can solve a problem without knowing it first. The full array and dimensions of the problems before us, and their degrees of urgency, are still imperfectly understood. Few scientists now are capable of viewing the system as a whole. The age of the poly-math is long over, and the age of the specialists presents grave difficulties, not the least of which is the tendency toward scientific myopia.

Somehow or other we must get together to coordinate disciplines and heretofore-separate national efforts. It is not enough to meet and talk occasionally, though that is certainly essential. We must continually try for a closer meeting of minds and points of view, from whose interaction a new synthesis of ideas and solutions may be born with hybrid vigor. Beyond that, we must work together more closely than before. We have by now identified, it seems, many of the separate pieces of our jigsaw puzzle and know. approximately where they belong. Now we must begin to fit them together systematically to describe the continum of a field, or of a greenhouse. In a sense, this is a reversal of the prior tendency to split the large problem into specific isolated components. Of particular interest is the important trend to devise closed systems of production to contain the field in an artificial environment in which such variables as temperature, light intensity and composition, atmospheric humidity and carbon dioxide content, and soil water and nutrient content will all be controlled, as will be all measures to combat pests and diseases. Such systems will not replace the open field all at once, and perhaps never completely, so many of our present-day field problems will probably remain with us for quite some time to come.

In the meanwhile, what should we investigate most urgently? This is a question which scientists ought to consider together. It is important that we

PROLOGUE

recognize the hidden problems which appear innocuous because of their minute effect at any instant of time, but which may build up slowly into a dangerous threat to the environment or to human welfare. Once a problem has been identified and assessed, a solution must be found, not just any kind of solution, but one which fits the social and economic realities of each country. This, too, requires coordination and cooperation.

The choice of a research problem was up to now left to the individual scientist's own hunch or interest. This freedom of choice must continue, for research ought not to be regimented. But we might offer help to individual scientists (on a voluntary basis to be sure!) by pooling our brain power in charting out necessary and promising directions. At present, too many competent research workers, for lack of imagination or of an overall view of the field, spend their entire careers repeating the kind of experiments they did for their doctor's degree. Too often, these experiments, though meticulously done, eventually become trivial and hence wasteful of our scientific potential. Clemenceau once made the remark that war is too important to be left to the generals. Let us not give cause for anyone to say that research is too important to be left to the direction of the scientists.

We must remember that science is a social creation as well as an individual creation. We as agricultural scientists have long tried to be responsive to the needs of society, and must continue in this tradition. No lesser a scientist than Albert Einstein once said: "It is not enough that we understand about applied science in order that the creations of our mind shall be a blessing and not a curse to mankind. Concern for man himself and his fate must always form the chief interest of all technical endeavors. Never forget this in the midst of your diagrams and equations!"

There are pessimists about who would have us believe that science and technology are in vain, the gods that failed. They proclaim that the applications of science and technology have done more harm than good, and that scientists tend to be amoral and ethically blind. Ehrlich in his book, "The Population Bomb," states categorically that the battle to feed humanity is over, and that hundreds of millions of people are doomed to die of starvation in spite of any crash program we might embark upon now. A recent editorial in a scientific journal even stated that the most threatening pollutant of the modern world is the human baby.... . Science and technology, say R. and L. T. Rienow in their recent book, "Moment in the Sun," while providing us with the refinements and glamour of modern living, have also caused foul pollution of our earth, water, and air; noise, strain, tension, mental aberrations, increased leukemia and thyroid growths, liver and degenerative diseases of the very young, deposition of radioactive materials and pesticides in the bodies of unborn babies, rocketing highway death toll, accidents and poisonings without number. Indeed, quite a damning list of accusations!

To be sure, all these ills are not the fault of science and technology per se, but of their misuse. Science is pure knowledge; it does not act of itself. Even technology is not really a living Golem or Frankenstein, merely a tool, no better than those who wield it. We cannot allow this wholesale condemnation of science any more than the extreme opposite, which is a smug and complacent assumption that science is infallible and will work out all our problems automatically. If the potential of our science is to be realized for the benefit of mankind, it must be by our humanistic concern and dedicated and responsible action in which heart and mind must be partners.

Scientists are inveterate optimists, constant believers that something positive can and should be done. Perhaps not all of us can echo the booming confidence of Rutherford: "This is the heroic age of science!" We tend to agree with C. P. Snow who stated with somewhat more humility: "Each of us is solitary; each of us dies alone; all right, that's a tragic fate against which we can't struggle − but there is plenty in mankind's condition which is not fate, and against which we are less than human if we do not struggle." We must struggle against the condition that most of our fellow human beings are underfed and die before their times and against the condition that our environment and that of our children is deteriorating. We in the field of agricultural science are in a better position to do something about this than most of our contemporaries. Our task is one of a profoundly exciting and vast potential. The contribution our studies can make to peace and welfare is now at last being recognized. Surely it was no coincidence that last year's Nobel Peace Prize went to an agricultural scientist!

Perhaps these tasks are too great and too formidable for us to discharge in our lifetime. I wish therefore to conclude with the words of the Hebrew sages who said:

> "Lo aleikha hamelakha ligmor velo atah ben-khorin lehibatel mimena."

> "It is not for you alone to complete the task, but neither are you free to evade it."

D. Hillel

CRITERIA FOR DETERMINING THE AIMS AND DIRECTIONS OF RESEARCH IN SOIL PHYSICS AND TECHNOLOGY

W.R. Gardner
The University of Wisconsin, U.S.A.

Let me preface my remarks about the aims and directions
of soil physics research with a statement that I do not
think it possible to write rules and instructions for the
performance of outstanding research. Greatness in any area
of human endeavor is achieved, in part, by inspired defi-
ance of rules and conventions. Nevertheless, there is much
in soil physics research today that appears mediocre to me
and that could be improved by attention to a few simple
principles. This will not be a comprehensive check list by
which administrators and project leaders can ensure effective research. Though I am not above oversimplification
and overstatement in order to make a point, I would like
these remarks to be considered in the proper context. It
is my judgment that the quality as well as the quantity of
soil physics research is as great as at any time in the
history of soil science. It is only because of the task
ahead of us that I choose to dwell upon weaknesses I see in
our research programs.

I cannot speak of other countries and other cultures,
but observation of two decades of soil physics research
in the U.S.A. has led me to believe that one of our weakest
points in the research process is in the selection of the
individual research project. In our educational institutions and our research laboratories we spend too little
time helping our young scientists develop the critical
judgment so necessary to the selection of a worthwhile project. In theory, a good thesis topic should make a good
research project. In practice this is too seldom true.
Our young people are often assigned a project in graduate
school with little understanding of the process by which
the professor decided that it was a suitable project for
research. Even many of our mature scientists have not
learned this lesson well, as shown by the speed with which

research fads move through the soil physics community. The most frequent weakness of manuscripts which I am asked to review is the initial plan of research. There is no way a reviewer can help such an author except to point out how he might have approached the problem in the first place.

I don't expect anyone to quarrel with the idea that every research project should have a goal. It is less commonly taken into account that every research career should (and almost always does) have a goal. In the case of basic research these goals tend to converge or overlap. Applied research implies that the goals and wishes of a wider constituency must be taken into account. Every scientist has strong and weak points, biases and prejudices. Most scientists worthy of the name try to take these into account as they conduct their experiments and analyse their data. Many of us tend to be much less objective during the time we are setting our research goals. Failure of a scientist or his organization to take the idiosyncracies of the individual into account often leaves his administrator disappointed and the research team frustrated.

If we must have a goal, what then should it be? I believe that L.A. Richards was absolutely correct when he stated repeatedly that we should attempt to define and measure the <u>physical</u> <u>properties</u> of soils. By physical properties he meant those properties that were independent of the method of measurement. I would add to this as a part of our goal the <u>understanding</u> of the physical processes in the soil and the <u>communication</u> of that understanding to our colleagues in other disciplines and to our successors.

How do we go about defining and achieving that goal?

If you examine the collected works of any really good soil physicist I am sure that you will find a development and growth which cannot be attributed to mere random chance. Since we seldom move directly towards our goals it is that much more important that goals exist and be recognized. Most of you will recall from your studies of statistical mechanics the random walk or "drunken sailor" problem. In brief, if a sailor falls each time he takes a step and, upon rising, forgets completely which direction he is going so that his next step is in a random direction, after N number of steps there is an equal probability that he will end up in any given direction. If you start with a shipload of such sailors you will end up with a large pile of them at the starting point. On the other hand, if the

sailor retains even the most faint memory of his original
direction, slow but perceptable progress in that direction
will result. Thus it is in science. We do not move unerr-
ingly towards the heights. We stumble often and move in
false directions. The only time most of us are on the Bib-
lical "straight and narrow path" is when we cross from one
side to the other.

A good scientist probably stumbles as often as his lesser colleagues for he more frequently attempts the treach-
erous paths. It is the alacrity with which he picks himself up and the remembrance of his intended direction that
moves him out ahead.

As penance for accepting the assignment of this paper I
read through the 1970 edition of the Water Resources Research Catalog published by the U.S. Department of Interior.
My feeling after completion of this task, in addition to
that of stupefaction, was that our stated goals are either
too grand or too puny. I also had the strong suspicion
that many of the justifications for the particular research
in question were conjured up after the fact, rather than
taken into account prior to, and during the process of de-
cision making. Many of us are vulnerable to the criticism
voiced by G.W. Millar, Director of Research, John Deere &
Co. (1969), who stated:

> A problem in converting university research
> effort...is the almost calcified approach of
> universities to multidisciplinary efforts....
> The chance of bringing to bear a team of
> agronomists, engineers, physicists, or aerody-
> namicists on the solution to a multidiscipli-
> nary problem is almost non-existent.... I
> would like to remind my audience that there is
> much more similarity in the actual world of
> nature than in the world of nature as inter-
> preted by man.... The basic principles of sci-
> ence are universal and it is only through our
> almost bureaucratic effort that we as human
> beings cause fragmentation.

We are not always honest about our goals, but they re-
veal their presence or absence in due time. Rene Dubos recently wrote (1970):

> ...academic scientists are becoming more
> and more oriented toward theoretical problems

3

which, even when they are directly relevant to agriculture, are not likely to exert a significant effect on its course. Yet, there are many reasons to believe that the directions of the agricultural enterprise will soon have to be altered, because some of its present practices are incompatible with ecological constraints.

...ecological instability is increasing at such an accelerated rate that we cannot delay much longer the development of a nearly "closed" system.... In my opinion, the nearly closed system...will soon compel a reorientation of the agricultural enterprise, in its theoretical as well as practical aspects. Such orientation cannot be successful without creative effort on the part of academic scientists and their willingness to refocus their efforts on the problems of the contemporary world.

To the extent that the above observation is true, it is a serious charge. Agricultural physics resulted early in notable successes such as those exploited in the development of irrigation and drainage systems. In recent years it has been chemistry and chemicals which have been prime factors, both in the solution of agricultural problems and the creation of environmental problems. Whether in the most or the least developed country, the time has come for those who best understand the physics of the biosphere to make their presence known and voices heard.

We must set as one of our goals the more effective communication of our present knowledge and concepts. Too few scientists who need to understand the physics of the soil have adequate access to this knowledge. Admittedly, it is difficult to communicate this knowledge and the process does not occur spontaneously. It was Max Planck (1936) who observed that

An important scientific innovation rarely makes its way by gradually winning over and converting its opponents: it rarely happens that Saul becomes Paul. What does happen is that its opponents gradually die out and that the growing generation is familiarized with the idea from the beginning.

How many of the growing generation of ecologists are

learning soil physics from the beginning? Biologists,
ecologists, geologists, economists, etc., ad infinitum are
writing and speaking as though they had just invented the
environment. If the work of the soil physicist is ignored
we will all be the poorer for it. We will also be to blame
for it.

We must resist the calcification of our research organ-
izations and refuse to accept arbitrary limits on our
studies when and where they hamper progress. I regret that
it is virtually a matter of official policy in the United
States that the hydrological cycle is not a cycle. Rather,
it is a process which has its start in Agency X, passes
through Agency Y and ends in Agency Z.

It is true that bureaucratic constraints often hamper
our research efforts and result in a fragmentation of research. Instead of fighting and minimizing this fragmentation we as individual scientists often contribute to it.
At a recent meeting of the American Society of Agronomy I
was depressed by the large number of partial or incomplete
experiments reported. In the current vernacular in the
U.S., each scientist "did his own thing." Each measured
the variable with which he had become identified and for
which he had developed the skills, but left unattended important factors which were more foreign to him. If several
workers had been combined into a single team to work on a
single experiment the cumulative effect would have been so
much greater than that which actually resulted. Here was a
clear case of inadequately defined goals.

Perhaps one of our problems in soil physics is that we
tend to be found singly, or in two's and three's, at any
one location. In our attempt to seek comfort through a
community of like-minded scientists at other locations, we
have over-compensated to the point that we are working to
impress rather than to instruct. The attitude of the gnos-
tic which we sometimes appear to adopt diminishes our ability to make our research results useful. Perhaps we have
difficulty making our concepts understandable to non-physi-
cists because we do not understand them well enough ourselves.

Another thing which we must do is to make a greater effort to encourage our more able scientists to work on the
difficult problems which are ready for solution and rele-
gate our more pedestrian workers to less demanding, though
still important, tasks. One problem we have in the United

States, though I don't suppose anyone else is guilty of it, is that we seem to give our most important applied problems to our least qualified scientists. I am not sure why this is so. Perhaps there is a corollary of Parkinson's law which says that the product of the probability of success and the importance of the problem tends to be constant. Thus we have talented scientists working on relatively safe projects because they don't have to waste valuable research time getting financial support, and poor scientists with little chance for real success who compensate by working on the very important problems of society.

If we can succeed in bringing just a few young and bright scientists into soil physics and if we make a real effort to communicate with the many other disciplines which impinge upon our own it will result in a tremendous increase in our creativity. Creativity is a commodity which is in very short supply in soil physics.

The history of soil physics research confirms the idea that creativity often comes out of bisociation. Bisociation is a term coined by Arthur Koestler (1964) to describe the creative reasoning provoked by the interaction of two different universes or matrices of discourse. An entertaining description of this process is given by J. Kestin (1970), in which he uses examples of humor to illustrate the principle. According to this concept a creative situation exists when the ideas from one sphere or matrix intersect with the ideas from another matrix.

I would go almost so far as to claim that most of the fruitful concepts we now hold in soil physics came forth in the mind of individuals who saw a relation between a physical concept first developed in some other area and some process or phenomenon observed in the soil. It is for this reason that I have misgivings about training research soil scientists exclusively in soils departments. If they do not have one foot in some other camp there is little probability that they will push beyond the frontier of their discipline. No matter how well they till the broken ground there is little hope they will see beyond the horizon to the virgin territories.

In his Nobel Lecture, Richard Feynman (1966) makes a strong argument for seeking out the unusual path:

> I think the problem is not to find the best or most efficient method for proceeding to a discovery, but to find any method at all....

Theories of the known which are described by
different physical ideas may be equivalent in
their predictions, and are hence scientifically
indistinguishable. However, they are not psy-
chologically identical when trying to move from
that base into the unknown. For different
views suggest different kinds of modifications
which might be made, and hence are not equivalent in the hypotheses one generates from
them....

If every individual student follows the
same current fashion...then the variety of hypotheses being generated is...limited. Perhaps rightly so, for the possibility for the
chance is high that the truth lies in the
fashionable direction. But, in the off chance
that it is in another direction...who will
find it?

How does this relate to the objectives of this symposium? An advance look at the diversity of the topics and approaches is encouraging. There is evidence of health and
vitality. But rather than dwell upon our successes I would
like to comment upon one of our very real problems. As my
friend and colleague Milton Fireman once put it: "We only
know how to solve the easy soil problems." More often than
not we settle for an explanation of what has happened rather than a prediction of what will happen. We have many
gaps in our understanding of the physics of the soil. The
hydrologist works in units of days and in areas of many
acres. Unless it's raining; then he changes his units to
cubic feet per second. The micrometeorologist works in
centimeters vertically and who knows what horizontally,
with 15 minute time averages. The plant physiologist is
concerned with microns and seconds and the plant breeder
talks of generations. The soil physicist is somewhere in
the middle and is happiest if he can cast his problem in
dimensionless form so that he can point out quite grandly
that his solution will apply to any unit, if only he knew
what it should be. I think that we must not leave the synthesis and integration of our knowledge to the ecologist
and the economist. We must assume much of the responsibil-
ity. I think we are fooling ourselves if we think this integration will come automatically out of our computers and
our equations. They are essential tools and aids, but they

are not a substitute for thinking. The words of Ernst Mach (1894) seem apt:

> Intelligible as it is...that the efforts of thinkers have always been bent upon the "reduction of all physical processes to the motions of atoms," it must yet be affirmed that this is a chimerical ideal. This ideal has often played an effective part in popular lectures, but in the workshop of the serious inquirer it has discharged scarcely the least function.

We have come some distance since Mach wrote these words in 1894 but concepts such as "soil structure," "soil tillage," "erodability," etc. are hidden so deeply in our theories of the movement of water and air in the soil that we cannot elicit them when needed. Nor can we yet cope with such important processes as soil genesis in a quantitative way. We are not in a position to replace soil classification concepts such as "poorly drained," or "suitable for most crops" with a set of physical parameters which will allow us to manage our soil environment intelligently. We have a long way yet to go and our problems seem to be accumulating more rapidly than their solutions. Even so, other than the very general goal which I have stated, I will beg the question implied by the title of this piece and offer no real advice on how to define and achieve our research goals in soil physics, for I am mindful of the words of the historian Samuel Eliot Morison (1965):

> America was discovered accidentally by a great seaman who was looking for something else; when discovered it was not wanted; and most of the exploration for the next fifty years was done in the hope of getting through or around it. America was named for a man who discovered no part of the New World. History is like that, very chancy.

May we all be so unfortunate!

References

1. Dubos, R. (1970). Quality of life: earth dependent. Agr. Sci. Rev. 8, No. 1 (Editorial).
2. Feynman, R. (1966). The Development of the Space-Time View of Quantum Electrodynamics. Nobel Foundation.
3. Kestin, J. (1970). Creativity in teaching and learning. Amer. Scientist 58, 250-257.
4. Koestler, A. (1964). "The Act of Creation." Hutschison, London.
5. Mach, E. (1894). On the Principle of the Conservation of Energy.
6. Millar, G.W. (1969). Objectives of industrial research. In "Research with a Mission." Amer. Soc. Agron. Pub. No. 14, 39-48.
7. Morison, S.E. (1965). "The Oxford History of the American people."
8. Planck, M. (1936). "The Philosophy of Physics."

EFFICIENT MANAGEMENT OF WATER IN AGRICULTURE

T.J. Marshall
Commonwealth Scientific and Industrial Research
Organization, Australia

There is an ever growing need to control and use the world's resource of fresh water as wisely as possible so that demands made on it by the growing human population can be met. Many countries have been confronted with water deficiences throughout their existence. Others, more liberally provided with natural water resources, are now finding difficulty in satisfying the increasing urban and rural demands. At the same time, quality of the available water is endangered, too, as expanding towns pollute water with wastes and as intensified agriculture makes its own contribution through salinity, pesticides, and nutrient enrichment in drainage and run-off waters.

In the water balance equation,

$$P + I = E + A + U + \Delta W \tag{1}$$

the various items, precipitation (P), irrigation (I), evapotranspiration (E), surface drainage (A), underground drainage (U), and change in soil storage (ΔW), are all subject to control by man in some degree. This paper covers some of the ways available for increasing the use of water in agriculture through soil and water management practices that affect one or other items in this equation. It deals more particularly with areas of deficient precipitation where additional water may be supplied by irrigation or where the items on the right hand side of the water balance equation may be manipulated for better yield of crops.

Irrigation Development

More than 200 million hectares of land are irrigated throughout the world according to Framji and Mahajan (1969) on the basis of statistics collected by the International Commission of Irrigation and Drainage. About one half of this land lies within two countries, China and India. Some of it is in arid areas where agriculture must depend wholly on irrigation, as in Egypt, and some is in areas where irrigation supplements a rainfall that is by itself sufficient to sustain agriculture in some form, as in northwestern Europe and the eastern part of the United States of America. The area under irrigation has doubled in the last 30 years and further development is assured by the many large dams being planned or under construction. The costs of these projects may not always be repaid by the economic benefits because in the future, as in the past, irrigation works will no doubt be undertaken for humanitarian or political as well as strictly economic reasons.

Great quantities of uncontrolled surface water are available in some locations. In the Amazon alone the water going to the ocean is almost enough to irrigate the present irrigated area of the world. Often, as in the Amazon, use of the water in or near the river basin is made impracticable by unsuitable land or climate and a few large transfers of water from one basin to another have been undertaken because of this. Water is transferred, for example, from the Sacramento to the San Joaquin valley in the United States and, in Australia, water from the Snowy Mountains that formerly flowed directly to the coast is now led inland to the Murray River system.

In a review of the future of irrigation, Framji and Mahajan (1969) predict that there are 500 million hectares of potentially irrigable land in which usable land, suitable climate, and controllable water are in reasonable coincidence. Similarly Kalinin and Bykov (1969) predict that, by the year 2000, irrigation will require 7,000 km^3 of water. These two predictions are reasonably compatible. Kalinin and Bykov hold the rather controversial view that with the large transfers foreseen in the U.S.S.R. and elsewhere, rainfall may within 50 years be lifted by 30 to 40 mm in the Afro-Eurasian land mass because of the increased evaporation there.

Irrigation Efficiency

Opportunities for close control of water are obviously greater under irrigation than under rain. Yet the many failures that have marked the long history of irrigation demonstrate that the artificial process of adding water

gives less assurance of agricultural permanence than the
natural process. The extra water added in irrigating up-
sets the equilibrium that the soil had long established
with its environment. In particular, subsoils may become
so persistently wet that soluble salts are transferred to
the surface soil in arid regions and crops and soils are
damaged. During the present century much has been learned
about how to manage irrigated land but nevertheless the efficiency of irrigation is often low and salinity remains a
problem (Kovda, Berg, and Hagan, 1967).

Irrigation efficiency can be expressed as the fraction
of the water withdrawn from river, reservoir, or aquifer
that is used in evapotranspiration on the farm. Two components of irrigation efficiency can be usefully distin-
guished: efficiency in conveying water to the farm and efficiency in applying it to the fields. Seepage, evaporation, and consumption of water by unwanted vegetation along
the way from the source to the farm will reduce the fraction delivered to the farm, and thus affect water conveyance efficiency. Run-off from the farm or excessive or uneven applications causing percolation to depths beyond the
reach of plant roots will lower the water application efficiency (the fraction of water delivered to the farm that is
used in evapotranspiration). In a study made by the U.S.
Bureau of Reclamation on 22 of their projects in the western states of the U.S.A. (Jensen, Swarner, and Phelan,
1967; Erie, 1968), average water conveyance and water application efficiencies were found to be 0.62 and 0.58 respectively (Table 1). The product of these two, representing the efficiency in taking water from the source of supply to the crop, was thus only 0.36.

Improvement of irrigation efficiency may be a matter of
relative costs; when water is scarce or expensive more care
will be exercised in conveying and applying it. Steps that
are taken to reduce conveyance losses include lining or
piping of the distribution system and control of vegetation
along unlined parts of the system. Within the farm, efficiency can be improved by reducing seepage from the farm
distribution system, by preventing or reusing run-off, and
by applying the required amount of water uniformly and at
the right time. In practice, high application efficiency
depends much on good lay-out and on selecting an irrigation
system suited to the soil, water and crop. Thus sprinkler
systems can distribute water more uniformly over sandy

13

soils than can furrow or border check systems but on the
other hand saline water can be more damaging to foliage under sprinklers.

TABLE 1
Amount of Water Diverted, Delivered, and Used
(in surface cm) in 22 Irrigation Projects Examined by
the U.S. Bureau of Reclamation.
(Adapted from Erie, 1968).

Diverted	Delivered to farm	E	P	Consumption of delivered water	Efficiency in conveying to farm	Efficiency in applying at farm
a	b	c	d	c-d	b/a	(c-d)/b
157	98	77	20	57	0.62	0.58

Excessive applications (causing damage to land and
waste of water) and insufficient or badly timed applications (causing lowered yield or quality) can be checked by
instruments for measuring soil water or by the widely used
formulae of Penman (1948) and others for estimating evapotranspiration from meteorological data. It is generally
assumed in applying these formulae to irrigation control
that availability of water for transpiration remains unchanged until the soil dries to a suction of about 14 bars.
However it is recognized for certain crops and certain
stages of growth that the drying should be limited to a
lower suction (of the order of 2 bars) for best production.

Drainage

The total area of land with artificial drainage is
about 100 million hectares according to Framji and Mahajan
(1969). Much of this is in humid areas. The United States
of America for example has 37 million ha. of drained land
which is twice as great as the whole area of irrigated land
in that country. The intensive use of drains in some humid
areas can be well illustrated by the Netherlands where, out
of 2.5 million ha. of cultivated land, 1.5 million are
drained (including 0.5 million with tile drains). Within
arid and semi-arid irrigated areas the general need for

drainage if agriculture is to be permanent is now widely recognized.

The main objectives in draining are to remove unwanted surface water and to control the depth to the water table, which should be deep enough to allow good root development in humid areas and to prevent salt rising in irrigated arid areas. The usual depth for drainage differs widely in the two cases. While 2 m may be necessary in arid areas, 1 m may be sufficient in humid (Ogrosky and Mockus, 1964). In places where sub-irrigation is practiced, the water table may be brought up to depths of less than 1 m during dry periods. This improves the use of water in permeable soils in humid areas such as the Netherlands and the Everglades of Florida, U.S.A. (Criddle and Kalisvaart, 1967).

The moisture profile for water moving upward from a water table to the root zone or to the evaporating zone is obtainable from the equation

$$z = -\int_0^\psi d\psi/(1 + q/K) \tag{2}$$

where z is the height above the water table, ψ is the matric potential, q is the rate of upward flow, and K is the hydraulic conductivity which varies with ψ. From this it is possible to calculate the moisture profiles that develop for different rates of upward flow. In this way Wind (1961) showed for humid areas how depth of water table affected the supply of water to the root zone and Gardner (1958) showed for arid areas how it influenced upward transport of salt. Their conclusions accord with .the practices outlined above.

Salinity of Soil and Water

With good natural drainage, soil salinity is usually readily reduced when irrigation is undertaken with good quality water. This may be illustrated by data of Kalinin (1969) from a field experiment on Cotton State Farm No. 1 in southern Kazakhstan, U.S.S.R. on the leaching of saline land. The soil had not previously been leached and about 0.2 m of the water was required to bring it to field capacity to a depth of 1 m. Leaching was carried out with separate applications of about 0.4 m of water several days apart. The results show that with amounts exceeding 1 m of water, salts were greatly reduced in the top 1 m of these soils.

TABLE 2
Salt Removed from Saline Soil by Leaching
(Kalinin, 1969)

Amount of water applied	Salt removed from top 1 meter of soil as fraction of amount initially present	
meters	Total dissolved solids	Chlorides
1.5 - 2	0.75	0.96
1.1 - 1.4	0.70	0.96
0.7 - 1.0	0.55	0.89
0.4 - 0.5	0.30	0.84

If neither natural nor artificial drainage is satisfactory, salinity is likely to be a problem whether or not the soil was originally saline in arid and semi arid areas. Salts brought in by irrigation water and remaining after it is transformed to water vapor will tend to accumulate if not removed in drainage. If, in the water balance equation (1), we take A and ΔW as zero it is possible to write as an equation for the salt balance

$$C_i I + C_p P = C_e E + C_d D \qquad (3)$$

where each C represents the salt concentration of a particular item in the water balance equation (irrigation, precipitation, evapotranspiration, or drainage) as indicated by the suffix. Since no salt is removed in evapotranspiration (except that taken away with the harvested crop which is disregarded) and since C_p may be taken as zero, equation (3) reduces to

$$D/I = C_i/C_d \qquad (4)$$

This represents the "leaching requirement" (Richards, 1954) which must hold if there is to be no change in the salinity of the soil. Combining equations (1) and (4) yields

$$I = (E - P) C_d/(C_d - C_i) \qquad (5)$$

which defines the amount of irrigation water required to

hold the soil at a constant salinity under given conditions of evaporation, precipitation and salinity of irrigation and drainage waters. The highest concentration that can be tolerated in the soil solution by the particular crop being grown is taken as the limiting permissible value of C_d.

Although this leaching requirement is real enough, excess water is so commonly used in irrigation practice that it is usually met automatically unless the soil is of low permeability or the subsoil drainage inadequate. However, from equation (5), it follows that the higher the salt concentration, C_i, of the water the greater is the amount needed to meet the leaching requirement.

The possibility of increasing the use of poor quality water in irrigation is being explored in a number of places. Sandy soils as used by Boyko (1968) in Israel, offer the best possibilities especially if exchangeable sodium is a hazard in the irrigation water. Forges (1970) reporting on work in Tunisia in an area with some natural rainfall states that crops can be raised and salinity controlled on a variety of soils using brackish water, but production is lower than with good water. In Italy, according to Cavazza (1969), 1% of the irrigated land is irrigated with brackish water with salt content up to 5,000 ppm. Management of the terra rossa soils used there includes frequent applications with matric suction kept below 1/3 bar, and restricting irrigation of annual crops such as canning tomatoes to one summer in every two. The amount of rain falling in the two winters between each irrigation season is often sufficient to prevent the soil deteriorating. Cavazza suggests that if there is not enough rain to do this, occasional use of desalinized water might prove practicable in conjunction with brackish water.

While it may not be difficult to leach well drained soil, the displaced salt may affect the salinity of the river or aquifer when drainage water goes into it. In the Murray River in Australia, as in other irrigated basins, this contributes to a progressive increase in salinity down stream (Table 3). To help control the salt entering the river, the water from artificial drainage systems is being increasingly disposed of in evaporation basins isolated from the river, except perhaps during an unusually high flood when the accumulated salt can be flushed away.

TABLE 3
The Mean Salinity of the Murray River, Australia,
in 1966-67
(Data from River Murray Commission, 1970)

Station	Distance up river from ocean km	Total dissolved solids parts per million
Below Hume Dam	2,200	37
Torrumbarry Weir	1,640	65
Boundary Bend	1,250	138
Red Cliffs	910	151
Berri	525	324
Tailem Bend	89	391

Control of Soil Water

Land can be managed in various ways for more efficient
use of its water. The most important ways of ensuring the
best return per unit of soil water consumed are by selection of suitable plants, use of fertilizers, and good control of weeds and pests. The simplest examples of this are
where wasteland and the wasted water falling on it have
been transformed into usable soil and water. Thus, after
the discovery of a copper deficiency, 2 million hectares of
useless scrubland in southern Australia were developed and
a volume of 9 km^3 of water falling on this land (under an
annual rainfall of about 450 mm) was put to use. In the
past we have looked on such transformations solely as a
saving of land; but the time has come when we·have to think
increasingly of the water that is being brought into use by
better management. This is an extreme case. For more usual circumstances Viets (1962) has reviewed evidence showing
that dry matter production per unit of water used can be
greatly increased if fertilizers increase the yield of irrigated soils. Similarly the high yielding varieties of
wheat and rice that have come from research groups in Mexi-
co and the Philippines can, when well supplied with fertilizers and water, return more per unit of water than the

varieties they are replacing.

A serious source of loss that is difficult to do much about is the evaporation of water directly from soil in fallows, from the bare soil between plants, and over that large area of the world on which rain falls in too small a quantity to support useful vegetation or to fill dams. Ways of concentrating thinly distributed rain water on to restricted areas have been developed and have come into limited use. Myers (1967) has experimented with asphalt, water-repellent materials, plastic sheets, and exchangeable sodium in the soil to increase run-off into dams in arid areas. By removing permeable top soil and suitably shaping the bared kaolinitic subsoil, water-supply engineers and farmers construct "roaded catchments" in the south-west of Australia (Public Works Department, Western Australia, 1956).

For crop production, fallowing remains the main method for conserving water in soil but this depends on preventing transpiration rather than controlling direct evaporation from soil. Weedicides are increasingly used partly for this purpose too. Artificial covers are coming in for increasing use over soil but only with highly intensive crops. In the research field the micro-shedding of rain water into crop rows (Willis, Haas, and Robins, 1963) and on to planted sites (Hillel, 1967) offers some promise.

The amount of rain water entering soil can be increased by creating a favorable structure, terracing and by other measures commonly used in water and soil conservation to reduce run-off and increase soil storage. Deep penetration of water to recharge aquifers can be encouraged by water spreading. On the other hand the penetration of water beyond the depth of usefulness to plants on sandy soils can be wasteful. Erickson, et al. (1968) have experimented with asphalt barriers at depths of about 70 cm to help hold back water in deep sands in Michigan, U.S.A. Others have added fine material to sandy soil to improve retention.

Underground drainage is an item of the water balance equation that is often difficult to determine and control in other than artificial drainage systems. Two examples of its importance may be cited. After the native perennial vegetation was removed during development of agricultural lands in Western Australia, less water was used by the annual crops and pastures that replaced it. The increased underground drainage to the valleys caused shallow water

tables to develop and 100,000 hectares of land and many water supply systems became useless because of salinity. The
second case concerns the recharge of ground water in a
shallow aquifer in South Australia. In a water balance
study of the region in which lysimeters and soil water instruments were used, Holmes and Colville (1968) obtained
data given in Table 4 showing that the amount of recharge
was substantial and that considerable development of irrigation from this resource could therefore be undertaken.
Recharge under pine forest in the same region appeared to
be negligible from data obtained with the neutron moisture
meter. This difference between forest and grassland is being studied further by measuring the natural abundance of
tritium and also by measuring the hydraulic gradients in
the ground water under grassland and forest.

TABLE 4
Drainage to the Water-Table Aquifer under Grassland in mm
(Holmes and Colville, 1968)

	1963	1964	1965
Precipitation	428	748	515
Underground drainage	40	134	72

Management in Relation to the Environment

With the rapid growth of the world's population, there
is continuous pressure to develop new land and intensify
the use of the old. Despite this we cannot afford to over-
look the hazards to the environment created in developing
land, harnessing water and increasing the use of fertilizers and pesticides. We are constantly reminded of these
hazards by damage already done in erosion of land, pollution of rivers, and salinity of both. About 40 years ago
soil erosion had reached a serious stage in those agricultural lands of the world that were developing rapidly at
that time and preventions and cures were made following an
alert sounded initially in the United States. We are now
approaching a greater crisis of environment that involves
air, land, and water more widely. Soil figures in this
crisis in various roles - as an effective sink for residues,
as a zone of biodegradation, as a source of pollutants in

drainage and run-off water, and as a system taking a great amount of water away from the streams and aquifers of the world wherever irrigation is practiced. As was the case with soil erosion, ways of curing pollution and maintaining environmental quality are fairly well understood, but if present trends in growth of population and agriculture continue they will become increasingly difficult to apply. Efficient management must take care of these problems if our environment is to remain productive, healthy and satisfying.

References

1. Boyko, H. (1968). Farming the desert. Science J. 4(5), 72-78.
2. Cavazza, L. (1969). Agronomic aspects of irrigation with brackish water in southern Italy. In "Value to Agriculture of High-Quality Water for Nuclear Desalination," pp. 219-224. Int. Atomic Energy Agency, Vienna.
3. Criddle, W.D. and Kalisvaart, C. (1967). Subirrigation systems. In "Irrigation of Agricultural Lands," pp. 905-921. Agronomy 11, Amer. Soc. Agron., Madison.
4. Erickson, A.E., Hansen, C.M., and Smucker, A.J.M. (1968). The influence of sub-surface asphalt barriers on the water properties and productivity of sand soils. Trans. 9th Int. Congr. Soil Sci., Adelaide 1, 331-337.
5. Erie, L.J. (1968). Management: a key to irrigation efficiency. Proc. Amer. Soc. Civ. Engs. 94(IR3), 285-293.
6. Forges, J.M. (1970). Research on the utilization of saline water for Tunisia. Nature and Resources, UNESCO 6, 2-6.
7. Framji, K.K. and Mahajan, I.K. (1969). "Irrigation and Drainage in the World. A Global Review." Int. Comm. on Irrigation and Drainage, New Delhi.
8. Gardner, W.R. (1958). Some steady state solutions of the unsaturated moisture flow equation with application to evaporation from a water table. Soil Sci. 85, 228-232.
9. Hillel, D. (1967). Run-off inducement in arid lands. Final Tech. Rept. to U.S. Dept. Agr. Hebrew Univ. of Jerusalem, and Volcani Inst. Agr. Res., Rehovot.
10. Holmes, J.W. and Colville, J.S. (1968). On the water balance of grassland and forest. Trans. 9th Int. Cong.

Soil Sci., Adelaide 1, 39-46.
11. Jensen, M.E., Swarner, L.R., and Phelan, J.T. (1967). Improving irrigation efficiencies. In "Irrigation of Agricultural Lands," pp. 1120-1142. Agronomy 11, Amer Soc. Agron., Madison, Wisc.
12. Kalinin, G.P. and Bykov, V.D. (1969). The world's water resources, present and future. Impact of Science on Society, UNESCO 19(2), 135-150.
13. Kalinin, Ya. D. (1968). Experiment in the leaching of saline land in southern Kazakhstan. Soviet Hydrology: Selected Papers, No. 2, 201-209.
14. Kowda, V.A., Berg, C. van den, and Hagan, R.M., eds. (1967). "International Source-Book on Irrigation and Drainage of Arid Lands." Draft edition FAO/UNESCO.
15. Myers, L.E. (1967). Recent advances in water harvesting. J. Soil and Water Conserv. 22, 95-97.
16. Ogrosky, H.O. and Mockus, V. (1964). Hydrology of agricultural lands. In "Handbook of Applied Hydrology" (Ven Te Chow, ed.), pp. 21-1 to 21-97. McGraw-Hill, New York.
17. Penman, H.L. (1948). Natural evaporation from open water, bare soil and grass. Proc. Roy. Soc. A 193, 120-145.
18. Public Works Department of Western Australia (1956). Roaded catchments for farm water supplies. Bul. 2393, Dept. Agr. W. Australia.
19. Richards, L.A., ed. (1954). "Diagnosis and Improvement of Saline and Alkali Soils." U.S. Dept. Agr. Handbook 60.
20. River Murray Commission (1970). "Murray Valley Salinity Investigations." Canberra, Australia.
21. Viets, F.G. (1962). Fertilizers and efficient use of water. Advan. Agron. 14, 223-264.
22. Willis, W.O., Haas, H.J., and Robins, J.S. (1963). Moisture conservation by surface and sub-surface barriers and soil configuration under semi-arid conditions Soil Sci. Soc. Amer. Proc. 27, 577-580.
23. Wind, G.P. (1961). Capillary rise and some applications of the theory of moisture movement in unsaturated soils. Inst. Land and Water Management Res., Tech. Bul. 22.

SOIL TEMPERATURE AND CROP GROWTH

C.H.M. van Bavel
Texas A&M University, U.S.A.

Introduction

Second only to the volume of research on the physics of unsaturated flow of soil water, theoretical and experimental studies of soil temperature regimes have been abundant over a period of many years. But a candid assessment of current agricultural practices leads us to conclude that the impact of this work has been negligible.

A case in point is that of the potential value of mulching. The practice of providing or maintaining a relatively thin surface layer of some suitable material on the soil surface has a significant effect upon soil temperature, as even a superficial theoretical analysis would predict. A survey of mulching experiments (Jacks et al., 1955), shows that the effect can be either a decrease or an increase of the average temperature. Also, it suggests that either result may under different circumstances enhance or depress crop growth. No wonder, then, that the results of mulching experiments seem irreconcilable and defiant of attempts to predict, on the basis of physical principles, what mulching will accomplish with regard to optimizing soil temperatures for crop growth in a given setting of climate, soil, crop and season. The only resort left to the inquiring agronomist is to proceed by trial and error, in the time-honored fashion.

In this assessment of the subject, we will attempt to explain why so much effort has borne so little fruit to date. Also, we will suggest a different approach to research concerning soil temperature as a crucial factor in physical crop ecology. These suggestions are made in the hope that theoretical and physical analysis can be used in solving the practical problems of biological engineering that must be faced in crop production.

Current State of Theoretical Concepts

The most complete and recent review of soil temperature physics is found in Van Wijk (1966). Here we find an analysis of the fundamental heat flow equation, and of many methods to solve it for given boundary conditions. Also, we find an analysis of the two thermal properties, heat capacity and thermal conductivity, and their relation to soil constitution and water content.

It must be stated that the analytical approach, as exemplified by the original work of Van Duin (1956), Van Wijk (1966, Chapter 5), and Lettau (1951, 1954) has limitations. It can deal only with simply stratified soil, or with instances where the thermal diffusivity of the soil is a simple function of depth. In this regard the situation is similar to the usefulness of analytical procedures in solving the flow equation for soil water dynamics.

Also, the surface boundary condition describing the net radiative energy input must be stated as a simplified, periodic function to yield solutions for the temperature regime at a given depth. We must ask how many practicing agronomists will believe that all these simplifying assumptions are realistic.

A considerable gap in the theory is that, given the surface energy input in the form of radiation, it cannot predict the allocation of energy among the three components of soil heat flow, evaporation, and sensible heat flow in the air. This is true even for the simplest surfaces. Only when evaporation is zero is the analysis possible (Lettau, 1954; Van Wijk, 1966, Chapter 8).

A singular stumbling block in the way of applying the classical theory is the fact that the "apparent" thermal conductivity or thermal diffusivity depends significantly upon temperature, soil water content, and pore size distribution. The problem was signalled by De Vries (1958), who also attempted to provide a theoretical approach for predicting the thermal constants. In point of fact, it appears necessary that for each soil situation the inter-relations of thermal properties, temperature, and water content must be determined by elaborate calculations. In practice we shall, therefore, have available a set of empirical curves or data sets, that cannot necessarily be represented by manageable mathematical relations. De Vries (In Van Wijk, 1966, Chapter 7) has shown how the calcula-

tions are to be made, and has stated that the estimated accuracy is from 5 to 10%, which should suffice for many purposes.

An encouraging development has been reported recently.
Numerical and computer-aided methods have been used for
some time to solve problems in soil water dynamics. These
methods can also be used in soil heat transfer and they are
particularly effective with the help of dynamic simulation
languages. Such computer languages specifically recognize
the changes over time that take place in a system and allow
solutions based upon numerical or tabular inputs, in contrast with methods that need functional expressions.
Wierenga and De Wit (1970) recently showed that, using such
a language (CSMP/IBM 360/65), soil temperatures can be easily calculated at any depth and time, given the value of
the temperature at one point and the dependence of thermal
properties on temperature as well as on time and position.
The procedure does not presuppose any specific mathematical
form of any relationship and it is quite simple to carry
out.

The numerical simulation approach is a very general and
powerful tool and it could be made to include such features
as soil layering and the effects of water content. But its
impact on the problem at hand goes beyond the mere physical
analysis of heat flow problems, as we will suggest further
on.

Agronomic Significance of Soil Temperature Regimes

Knowledge of the role of soil temperature in crop
growth and development under field conditions, and the
methods used in its investigation, stand in sharp contrast
to the sophistication of current physical concepts. In experimental studies, usually, a series of treatments is involved that can be expected to affect soil temperature.
The latter is measured, typically, at no more than one or
two depths, sometimes periodically, more generally once or
twice daily. Measurements or estimates of soil thermal
properties are almost never made. Plant or crop behavior
is documented in terms of yield, development, and related
behavioral parameters that represent the accumulated effect
of plant processes over long periods of time.

The local and practical value of such experiments can
be considerable, but it is difficult to generalize the

findings to include other soil types and climatic conditions than those of the specific experiment. A typical and recent example of an agronomic study is one reported by Adams (1970) in which the effect of plastic films and petroleum mulches on the growth and yield of grain sorghum and corn was measured. Soil temperatures were measured at eight depths, hourly throughout the season, an unusually complete record thus being available.

However, the analysis was confined to average values at a single depth (7.6 cm), for obvious reasons. There simply is no known and rational procedure for summarizing or integrating the joint and total effect of what is occurring simultaneously at several depths on an hourly basis.

Adams concluded that the increase in soil temperature during the early season obtained by mulching treatments had a significant effect upon the rate of crop development. The sorghum yields were also increased, but not the yield of corn. In either case a definite cause and effect relation could not be established. A significant correlation was obtained between plant height and soil temperatures 10 to 20 days after planting.

It is possible to cite many similar field studies, as, for example, in the review by Jacks (1955). Typically, a good interpretation of the data is not possible because there exists no method to relate the findings to relevant data obtained in studies of crop physiology.

It must be made clear that the problem does not reside in a lack of fundamental information. In an excellent review dating back to 1952, Hagan (In Shaw, 1952, Chapter 5) listed many data on the effect of temperature on specific aspects of crop growth and root physiology. Significant examples are: the effect of temperature upon translocation and respiration, upon mineral absorption by root systems, and upon the rate of water absorption by roots. Root growth, measured as elongation of individual cells and of the entire organ, has been found sensitive to temperature, as well as the more complicated process of seed germination and sprouting of underground organs. Later work was added to this considerable body of knowledge, as evidenced by a review made in 1966 by Nielsen and Humphries.

A recent finding of considerable interest was reported by Walker in 1969 and 1970. In the first study, the functioning of corn seedlings was shown to be remarkably sensitive to root environment temperature, while constant conditions were maintained for the shoots. It should be recorded that the light levels in this work were unrealistically low, about 15% of maximum daylight. The recorded responses in growth and nutrient uptake were sometimes as much as 30 to 40% per degree C.

In the 1970 study, Walker found that alternating soil temperatures had a quite different effect than constant soil temperatures. Stripped from all the details, this finding simply demonstrates that the "average" temperature does not result in an "averaged" behavior. This must be so, inasmuch as the temperature response is not linear, as Walker showed in the 1969 work (see Figure 1). Also the temperature response is different for different plant functions or growth aspects.

Thus, we find here a convincing reason why crop behavior in a field cannot be related to average soil temperatures with any great degree of success. A parallel situation exists in crop ecology with regard to light measurements and crop productivity, as pointed out by McCree and Loomis (1969).

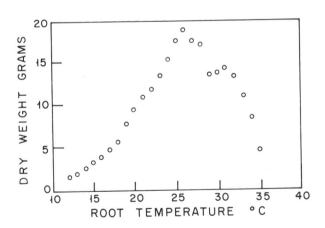

Figure 1. Total dry weight of 23-day old corn plants at different root temperatures (after Walker, 1969).

Relating Crop Response to the Soil Temperature Regime

It is difficult to see how any future progress can result merely from more effort along the following three established lines: (1) ever-more refined analysis of the heat flow problem in porous three-phase media, (2) more studies of the temperature sensitivity of specific plant functions made under controlled conditions, (3) further studies of crop response to treatments that affect the natural soil temperature regime.

Clearly, a synthesis is required and the main purpose of this paper is to suggest the method of such an approach. It is derived from the same principles that have led to models of photosynthesis by crop canopies. In these, as exemplified by the work of De Wit (1965), and of Ross (1970), the total effect of a continuously varying light regime upon photosynthesis is found by considering each canopy element singly, as well as the anticipated response, usually non-linear, of that element to the quantity of light absorbed. Then, the rates of photosynthesis are integrated over all hours of the day and all canopy elements to come to an aggregate estimated photosynthetic activity for the crop canopy for a 24 hour or longer period. It is obvious that such a procedure must be numerical to be practical.

Light interception models are not a total solution to the problem of crop growth, but a first step in this direction. A similar approach can be formulated for the problem of soil temperature and root system, or whole plant response.

We already have seen that it is now possible to generate the soil temperature at any point in time and space, given a series of observations at one point - inclusive of the surface temperature. If the temperature response of the root system is known, the calculated regime can be used to estimate the instantaneous root system response. The integral of this quantity over depth and time may constitute a worthwhile model to discriminate one treatment from another.

We propose that such a calculation should precede and, indeed, be the basis of field experiments that would be used as a test to find whether or not the modelling approach is correct and to what degree. Continuous system modelling languages seem to be well adapted for such work

6 9913

since the root response function and the depth distribution of root activity are generally available only as empirical sets of behavioral data. The peculiarity that the system response changes continually with time and position can be handled without difficulty and it could even be made to include growth or root proliferation rates.

We lack so far a specific example of this approach, but an experiment was reported recently by Wanjura et al. (1970) that, in a very simple form, illustrates what may be achieved.

In an effort to predict the time of emergence of cotton seedlings, use was made of soil temperature measurements at 5 cm depth every hour during the emergence period. These temperatures were used to compute the hourly elongation rates from work done by Arndt (1945) under controlled conditions (see Figure 2). These rates were integrated over time to find the time required to attain a total length of 5.0 cm, in each of a series of treatments.

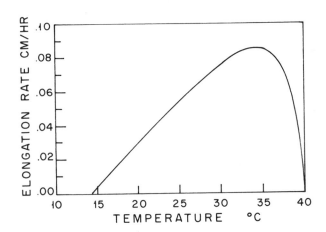

Figure 2. Cotton hypocotyl elongation rate as affected by temperature (from Arndt, 1945).

The results are summarized in Figure 3. On the average, a total emergence time of about 8 days was predicted with an accuracy of 3-4 hours. It is worth quoting the authors, where they conclude: "The modelling attempt discussed here is simple, but demonstrates that in many instances sufficient information is available to develop simulations of biological systems."

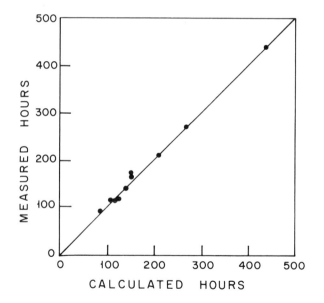

Figure 3. Calculated versus measured emergence time for cotton seedlings (from Wanjura et al., 1970).

Conclusions

We submit that the successful approach to simulation of the physical process of heat transfer in soils can be expanded. In this expansion, available information on the temperature sensitivity of root system functions is used to model the effect of treatment or of weather variables upon the response over time of the plant root system.

The merit of this suggestion will rest upon the outcome of properly designed crop growth experiments. Such experiments must reflect a joint planning by soil physicists, crop physiologists and agronomists. It would appear an advantage if such experiments were carried out in a controlled environment that was large enough to permit normal crop development and yield evaluation, and in which it should be feasible to simulate natural conditions. Both features are not yet available to agronomic research. The conventional phytotron, almost needless to say, is wholly inadequate for such work (van Bavel, 1971), whereas the field environment is uncontrollable, though still preferable.

The problem of disposing the reject heat from electric power plants is assuming serious dimensions in the industrial parts of the world. It has resulted in a new upsurge of interest in soil heating and in the potential value of the reject or process heat for promoting crop productivity. Possibly, the course of action that is outlined here may prevent much needless empirical research occasioned by the "thermal pollution" concern. Rather, it should prove possible to combine data already available from the literature to arrive at a preliminary assessment of the need for supplemental soil heat and the associated cost-to-benefit ratios.

In conclusion, it is appropriate to quote B.T. Shaw (1952) from his epilogue to the ASA monograph on soil physical conditions and plant growth. He pointed out that many physical, chemical and biological conditions influence plant growth simultaneously, and that only a fundamental, rather than an empirical, approach is practical in the long run. Realizing the complexity of the ensuing calculations, he suggested that "the new electronic calculating machines offer us hope." At this time, some twenty years later, these words seem to have been prophetic.

References

1. Adams, J.E. (1970). Effect of mulches and bed configuration. II. Soil temperature and growth and yield responses of grain sorghum and corn. Agron. J. 62, 785-790.
2. Arndt, C.M. (1945). Temperature-growth relationships of the roots and hypocotyls of cotton seedlings. Plant Physiol. 20, 200-220.
3. De Vries, D.A. (1958). Simultaneous transfer of heat and moisture in porous media. Trans. Amer. Geophys. Union 39, 909-916.
4. De Wit, C.T. (1965). Photosynthesis of leaf canopies. Agr. Res. Rept. 663, 1-57.
5. Jacks, G.V., Brind, W.B., and Smith, R. (1955). Mulching. Commonw. Agr. Bur, Tech. Commun. 49.
6. Lettau, H. (1951). Theory of surface temperature and heat-transfer oscillations near a level ground surface. Trans. Amer. Geophys. Union 32, 189-200.
7. Lettau, H. (1954). Improved models of thermal diffusion in the soil. Trans. Amer. Geophys. Union 35, 121-132.
8. McCree, K.J. and Loomis, R.S. (1969). Photosynthesis in fluctuating light. Ecology 50, 422-428.
9. Nielsen, K.F. and Humphries, E.C. (1966). Effects of root temperature on plant growth. Soils and Fertilizers 29, 1-7.
10. Ross, T. (1970). Mathematical models of photosynthesis in a plant stand. In "Prediction and Measurement of Photosynthetic Productivity." Proc. IBP/PP Trebon Meeting, PUDOC, Wageningen.
11. Shaw, B.T., (ed.)(1952). "Soil Physical Conditions and Plant Growth." Academic Press, New York.
12. Van Duin, R.H.A. (1956). On the influence of tillage on conduction of heat, diffusion of air and infiltration of water in soil. Versl. Landbouwk. Onderz. 62, 82 pp.
13. Van Wijk, W.R., (ed.)(1966). "Physics of Plant Environment," 2nd Ed. North-Holland Publ. Co., Amsterdam.
14. Walker, J.M. (1969). One-degree increments in soil temperature affect maize seedling behavior. Soil Sci. Soc. Amer. Proc. 33, 729-736.
15. Walker, J.M. (1970). Effect of alternating versus constant soil temperatures on maize seedling growth. Soil

Sci. Soc. Amer. Proc. 34, 889–892.
16. Wanjura, D.F., Buxton, D.R., and Stapleton, H.N. (1970). A temperature model for predicting initial cotton emergence. Agron. J. 62, 741–743.
17. Wieringa, P. and De Wit, C.T. (1970). Simulation of heat transfer in soils. Soil Sci. Soc. Amer. Proc. 34, 845–848.

IMPROVING THE WATER PROPERTIES OF SAND SOIL

A.E. Erickson
Michigan State University, U.S.A.

Increased population and the demand for more food has placed a strain on our soil resources in many areas of the world. Sand soils, infertile and droughty, are usually avoided because of low yields. With the asphalt moisture barrier technique, these droughty soils may be converted to productive soil with a modest capital input (Erickson, et al., 1968a; Hansen and Erickson, 1969; Anon., 1971).

Well-drained sand soils have a durable surface which can withstand the abuses of tillage and farming, have high infiltration capabilities allowing for quick intake of rainfall or irrigation water, good aeration characteristics which favor root development, and self-mulching properties which reduce the loss of water by evaporation. The major deficiencies of these soils is their low capacity for retaining water, requiring frequent rains or irrigations to keep them productive. The high subsoil permeability can make it difficult to distribute water uniformly. When furrow or border irrigation is practiced, large quantities of water are wasted due to excessively deep percolation on the application side of the field. Sand soils also have low cation exchange capacity and are subject to leaching losses of nutrients resulting in low fertility.

A properly engineered moisture barrier, such as an asphalt layer placed at some depth within the soil, can overcome the physical disadvantages of sand soils while retaining their advantages. An asphalt moisture barrier can double the retention of water in the plant root zone and thus prevent excessive deep percolation and consequent waste of irrigation water. The barriered soils can reduce the leaching losses by its greater storage capacity and can also restrict upward movement of salt from lower horizons. The barriered soil will retain its self-mulching characteristics, its high infiltrability in the surface soil, its surface durability and its favorable aeration characterisistics. Barriered sand soils can thus become optimal plant growth media. The inherent nutrient deficiencies of these soils can be overcome by use of fertilizers.

Asphalt moisture barriers are continuous films of asphalt with gaps every two hundred feet or so for drainage (Hansen and Erickson, 1969). After a heavy rain or irrigation the water percolates down to the barrier and then across to its edge. During drainage, a high water content and a high capillary conductivity prevail above the barrier and water can move rapidly to the edge of the barrier. At the edge, similarly, there is a high capillary conductivity because of the perched free water being released. There is also a higher potential gradient which is partially due to gravity. As drainage of the excess water from the barrier proceeds, all of these values decrease. When all of the free water has drained from the barrier, the capillary conductivity at the barrier surface decreases, as does the potential gradient. Just off the edge of the barrier, with the source of free water removed, the water rapidly drains by gravity and the capillary conductivity soon decreases. These phenomena cause low tension water to be detained over the barrier surface. There is an edge effect, extending for a meter from the edge of the barrier, which can reduce the amount of water held there by about one-half.

To illustrate the amount of water that can be perched above the moisture barrier, reference is made to Figure 1 (Erickson et al., 1968a), a series of capillary rise and drainage experiments made on an Ottawa fine sand with 93% sand (71% of which is in the 1/10 to 1/4 mm range of particle sizes). The two-day capillary rise in this soil amounted to about 55 cm. When the soil was first saturated and then allowed to drain to a free water surface for two days, 10% water by volume was found above the 70 cm height and about 35% water by volume at the 45 cm height. Water added to the top of a capillary rise column generated the low branch of the hysteresis loop. Figure 2 shows soil moisture suction values on the same sand soil in the field, with and without an asphalt moisture barrier, taken after a rain or after a 6-inch irrigation. After the rain, the soil drained to 74 millibars of suction in 24 hours and to 90 millibars in 6 days. The water retained on the barrier was at a suction of 34 millibars after 1 day, 35 after 2 days and 40 after 6 days. The 6-inch irrigation which initially flooded the barrier indicated a suction of 6 millibars- after 1 day, 15 millibars after 2 days and 25 after 6 days. In this field, the barrier was placed 60 cm below the soil surface.

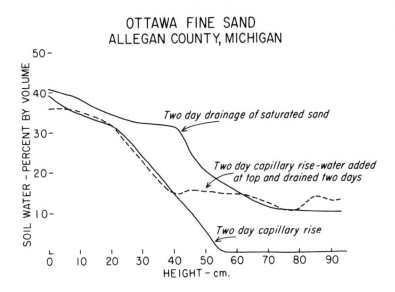

OTTAWA FINE SAND
ALLEGAN COUNTY, MICHIGAN

Figure 1. Water content vs height of capillary rise, unsaturated drainage and saturated drainage curves of an Ottawa fine sand.

Since the barrier was found to drain to a 25 millibar suction within 6 days after a saturating irrigation, as indicated in Figure 1, the soil values apparently operate between 25 and 85 millibars of suction above the barrier. The 20 cm of soil immediately above the barrier will thus contain over 30% of water by volume. This is in contrast to the 10 to 12% it would contain without the barrier. Assuming that 50% of the water held by sand soils is not readily available to plants, this means that a barriered profile can provide 200% more water to plants than a profile of sand without a barrier. Even if we consider the poorest case, in which water drains down to the original capillary rise curve, there would still be a 50% increase in the amount of readily available water in the barriered soil. The barriered sand soil thus has a capacity to hold water equal to that of loam soils. Furthermore, the extra water is held at low tension, which is probably better for most crops.

The fact that the barriered soil drains its free water in less than 24 hours would indicate that at least the upper 20 cm of the soil would be well-aerated at this time. This is usually sufficient for most crops. However, for deep rooted crops it might be necessary to place the barrier deeper in the profile or have the drainage gaps closer together to facilitate more rapid drainage of the excess water. Oxygen diffusion measurements made on sand soils with barriers have indicated that if the barrier is at a depth which allows for the upper 10 cm to approach the free drainage water content of the natural soil, there are no aeration problems. If the barrier is too shallow, aeration problems can exist for short periods of time during drainage.

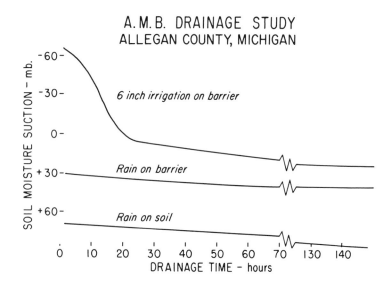

Figure 2. Field drainage curves of rain and irrigation water from an Ottawa fine sand with and without asphalt moisture barriers.

Placing the barrier at a depth that permits the surface
5 to 10 cm to drain freely also allows for the natural
self-mulching of sand to occur on the surface. This effect
helps preserve the extra stored water from evaporation. If
the barrier is too shallow so that the surface soil remains
moist and does not self-mulch, the extra water stored above
the barrier will be subject to evaporation through the surface.

The increased capacity of these soils to retain water
through the growing season has been verified by soil moisture measurements made with a neutron surface-moisture and
depth moisture probes. The results are shown in Figure 3
(Erickson et al., 1968a). Because of the depth at which
the barrier was placed, there is little difference in the
upper 15 cm of soil during the season regardless of treatment. In the 30 to 60 cm depth, however, more than twice
as much water is held in the barriered sand as compared to
the natural sand, either irrigated or unirrigated.

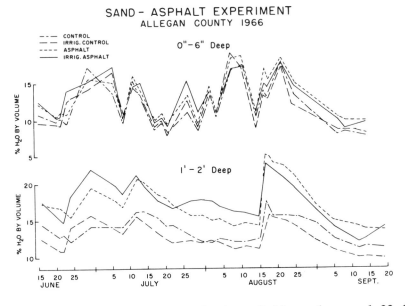

Figure 3. Soil water variations 0-15 cm deep and 30-60
cm deep during the growing season on asphalt barrier experimental plots of Ottawa fine sand.

The additional water retained in a barriered sand soil, which gives it a storage capacity equivalent to that of a better agricultural soil in the humid region, can produce equivalent yields if proper fertilization is applied (Erickson et al., 1968a; Hammond et al., 1967; Saxena et al., 1967). In irrigated regions, the barriered soil can also be expected to require less frequent irrigations.

Asphalt moisture barriers have been furrow irrigated both in Taiwan (Erickson et al., 1968b) and in Arizona very successfully. The barrier is brought to the surface at the irrigation ditch and thus deep percolation of water is eliminated. This has meant a saving of 50% or more of the irrigation water compared to the same sand without barrier. The additional water retained in the soil reduces the frequency of irrigation and some of the rainfall that would be lost to drainage in the natural soil is saved also. Asphalt moisture barriers allow furrow or border irrigation on sand soils which could not be efficiently irrigated by these methods otherwise.

Barriers have been used for restricting the movement of salt in the profile. In Taiwan, an experiment has demonstrated how the salt can be kept from moving up on polder soils during the dry winter season. This could probably be adapted in desert regions to keep the salt down during the non-irrigated season.

Asphalt moisture barriers have been field tested throughout the United States and various other countries (Anon., 1970; Erickson et al., 1968a, 1968b; Saxena et al., 1967). Sand soils with asphalt barriers can produce yields of crops equivalent to those of the best soils in their areas and much higher than those obtainable from the original sand soil. The difference in yields between barriered or non-barriered soils will generally depend on the drought stress during the particular season. Often the barriered soil gives higher yields than the best irrigation management possible without a barrier, since many plants apparently grow better with the low tension water in a barriered soil.

References

1. Anon. (1970). The Asphalt Moisture Barrier. Bul. published by American Oil Company, Whiting, Indiana.
2. Anon. (1971). Productive Soil from Sand. Bul. pub-

lished by Amoco Moisture Barrier Company, 910 S. Michigan Ave., Chicago.
3. Erickson, A.E., Hansen, C.M., and Smucker, A.J.M.
   (1968a). The influence of subsurface asphalt moisture
   barriers on the water properties and the productivity
   of sand soils. Trans. 9th Int. Congr. Soil Sci., Adelaide 1, 331-337.
4. Erickson, A.E., Hansen, C.M., Smucker, A.J.M., Li, K.,
   Hsi, L., and Wang, T. (1968b). Subsurface asphalt barriers for the improvement of sugarcane production and
   the conservation of water on sand soil. Proc. 13th
   Int. Soc. Sugar Cane Tech., Taipei, 787-792.
5. Hammond, L.C., Lundy, H.W., and Saxena, G.K. (1967).
   Influence of underground asphalt barriers on water retention and movement in Lakeland fine sand. Soil and
   Crop Sci. Soc. of Florida Proc. 27, 11-19.
6. Hansen, C.M. and Erickson, A.E. (1969). Use of asphalt
   to increase water-holding capacity of droughty sand
   soils. I and EC Product Res. and Dvlpmnt. Proc. 8,
   256-259.
7. Saxena, G.K., Hammond, L.C., and Lundy, H.W. (1967).
   Response of several vegetable crops to underground asphalt moisture barrier in Lakeland fine sand. Florida
   State Hort. Soc. Proc. 80, 211-217.

IMPROVEMENT OF SOIL STRUCTURE BY CHEMICAL MEANS

M. De Boodt
Rijksuniversiteit, Gent, Belgium

Introduction

The important influences of soil physical properties on plant growth have been recognized by some of the earliest research workers. Russell (1950) and Hénin et al. (1960) cited contributions by G. Markham in 1625 and G. Heuzé in 1882. However, when in 1952 the monograph on "Soil Physical Conditions and Plant Growth" (B.T. Shaw, editor) was written by a group of outstanding scholars, the contributors pointed out in the introduction:

> There is a widespread popular acceptance of the importance of the physical properties of soil to plant growth, but a large proportion of the statements commonly made on this subject are vague, qualitatively and frequently unsupported by factual evidence.

The principal reasons why a precise insight was lacking at that time and why an earlier start could not be given for appropriate research of the phenomena involved are:

a) Attempts were made to look at each soil physical growth factor separately, although means and tools did not allow this. Attempts to change only one factor, say soil structure, in fact often resulted in a change of other factors as well, e.g. the nutrient status of the soil. Applying natural organic matter was at the time a favored tool for modifying soil structure (Nielsen, 1963).

b) To detect the influence of each soil physical growth factor one must have the possibility of considering a multitude of primary and perhaps secondary interactions of the various factors involved. The assessment of some of these interactions became possible only with the advent of high-speed computers.

Other reasons, important to some extent, are that quite a lot of study was necessary to realize the multitude of circumstances which can lead to the same final result in respect to the interacting effect of a given physical factor on plant growth (e.g. different moisture regimes can result under different circumstances in the same plant development). As if all this were not sufficiently complicated, other interactions with vital plant growth factors are still waiting to be fully explored, e.g. the interdependence of soil physical factors on the availability of nutrient element (Grable, 1966; Nielsen, 1963) and on microbiological activity (Harris, Chester, and Allen, 1966). Fortunately, techniques are gradually becoming available for the precise study of phenomena formerly recognized only in qualitative terms.

Soil Physical Growth Factors

At present, when speaking about soil physical factors influencing plant growth, one thinks of four main items:
  a) Soil moisture;
  b) Soil aeration;
  c) Soil temperature;
  d) Soil mechanical properties.
To be complete, one must also add such physico-chemical factors as cation exchange capacity and salt content. An awareness of all these factors is needed for a reliable assessment of what can and should be optimized physically in order to achieve better plant growth.
It is not within the scope of this discussion to provide a detailed review of what has been achieved in better understanding the influence of each factor mentioned or its interdependence with other factors influencing plant growth. For these aspects reference is made to recent books treating the subject in a comprehensive way such as the ones by Bear (1965), Black (1968), and Hillel (1971). More specialized on the water relationships of plant growth is the work by Slayter (1967). Specific information on water deficits and plant growth is given in the two volumes edited by Kozlowski (1968). For more information on the behavior of different plants in respect to water availability at different growth stages reference is made to Salter and Goode (1967). The interactions of soil aeration with other soil physical factors and their combined effect on plant

growth were elucidated by Grable (1966). In recent years more refined techniques have allowed more precise control of the two main growth factors of soil moisture and aeration (Rawlins, 1971).

This preliminary work was necessary before effective steps could be taken toward modification and optimization of the soil physical environment.

Modification of Soil Physical Conditions

Attempts at changing some important aspects of the soil physical environment are not new. For thousands of years, practices such as irrigation, runoff inducement, and soil tillage have been known in the Near and Middle East. Other practices such as the application of artificial soil conditioners are fairly new. Consequently optimizing the soil physical environment can be carried out in a great number of ways. For the discussion here they will be subdivided into two groups: the so-called classical treatments, and the new ones based on the use of soil-conditioning chemicals.

To the first group belong: irrigation, drainage, soil mechanical management, amending soil with natural organic matter or with divalent salts (such as gypsum or magnesium and calcium carbonate). It is not our intention to discuss the classical treatments extensively, as these have been adequately treated in many previous symposia (see for example the proceedings of the 1968 Symposium on Problems of Soil Cultivation). This group of treatments has resulted in some remarkable achievements which are now coming into widespread practice. A few examples should be mentioned: irrigation through continuous water application (Hillel, 1971), and deep plowing (Netherlands Journal of Agricultural Science, Special Issue, 1963).

The new soil-conditioning treatments will be discussed here in some detail, as an impressive number of applications are now becoming economically feasible. Ever since such products were first introduced in the early 1950's, their number has increased and their range of application has diversified. Soil conditioners are no longer applied solely to induce flocculation of clay and stabilization of aggregates, but also for the many other physical characteristics they can impart to the treated soils. The new soil conditioners can be classified accordingly (De Boodt, 1970):

45

a) Products making the soil more hydrophilic (e.g., polyacrylamide solutions).
b) Products making the soil more hydrophobic (e.g., slowly breaking bitumenous emulsions).
c) Products for increasing soil surface temperature (e.g., mulches of fast-breaking bitumenous emulsions).
d) Products to stabilize the structure of tilled soil and hence to render it more penetrable by plant roots.
e) Products tending to increase the cation exchange capacity (e.g., emulsions with strong acid function, or aluminum magnesium silicate solutions, or zeolites).

The foregoing does not imply that the soil structurization and stabilization aspects of soil conditioning have been forgotten. These aspects are still essential except perhaps when mulches are applied.[1]

The important shift in emphasis regarding the mode of action of soil conditioners (first launched twenty years ago) which has lately taken place should be explained further. Whereas formerly flocculation of the clay particles was considered to be the essential function of the soil conditioner (Harris, Chester, and Allen, 1966), now the main stress is put on bond formation between the sand and clay domains (De Boodt, 1970). It is now clear that the clay particles are generally sufficiently flocculated in most agricultural soils to have clay domains present (Emerson, 1959).[2] What is lacking in many agricultural soils are bonds between sandy particles as well as between clay domains and sandy particles. Such bonds are needed to produce stable aggregates. To create such bonds in the right place, emulsion-based soil conditioners have been developed (De Boodt and De Bisschop, 1970; De Boodt et al., 1971). Some of the new soil conditioners can also act as flocculating agents.

It has often been observed that even soils which are adequately aggregated and stabilized may lack certain physical attributes to serve as optimal media for crop production. Some of these soils can be improved by using only

---

[1] In the present context, mulches are not considered as being real soil conditioners.

[2] Alkali soils, of course, are exceptions to this rule.

one soil conditioner, providing the right diagnosis and choice of treatment have been made. Let us cite a few examples:

  a) Sandy soils are in general characterized by low water retention and cation exchange capacity (C.E.C.) and high impedance to root development. With proper structurization and stabilization, these shortcomings can be rectified to some extent. The use of an emulsion with strong acid groups (e.g., HSO_3 on the micelles) can improve both the C.E.C. and the water holding capacity because water molecules will be oriented and attracted by each of the active sites. Experiments have shown that in this way available water can be increased from 1.5 to 6% by mass and the C.E.C. from 0.5 to 7 milliequivalents per 100 grams. Such increases will often result in better plant growth.

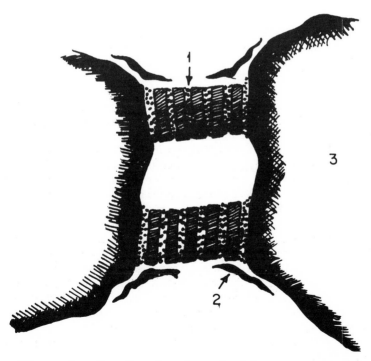

Figure 1. 1: clay domains, 2: bitumenous micelle, 3: sandy particle.

b) Heavy clay soils often indicate a tendency to poor
aeration and to salt accumulation. When such soils
are treated with an hydrophobic emulsion, the aggregates formed will still retain enough water, since
the insides of the clay domains are not touched by
the emulsion while at the outside the clay domains
are held together by long micelles which have the
effect of reducing the effective C.E.C. by up to 50%,
depending on the concentration of the soil conditioner used (see Figure 1). The salt content, as
measured in the 1:5 extract solution, is reduced in
the same ratio, as the micelles of the conditioner
fixed on the outside of the aggregates do not allow
the clay domains to swell and hence the fixed salt
or ions tend to escape into the solution.

Solutions of Polymers as Soil Conditioners

Interest in soil conditioners using artificial products
is not new. We recall the American Association for the Ad-
vancement of Science symposium in Philadelphia in December,
1951, where krilium was first introduced, and the special
issue of "Soil Science" in 1952 which was devoted completely to that subject. The introduction of that early soil
conditioner indicated how soil could be structurized and
stabilized at the same time, using a sodium salt of poly-
acrylonitrile. In less than two years, more than 100 pat-
ents were introduced for similar products. However, in
spite of much advertising, the success of these products
did not last. Although it was only necessary to treat the
soil with 0.1% of the soil conditioners, the cost of the
treatment was prohibitive (i.e. over $2,200 per hectare of
treated soil).

Notwithstanding the apparent initial failure, the concept of soil conditioning was launched and researchers continued to seek better and cheaper means for improvement of
soil physical conditions. This research has lead to the
selection of at least seven promising polymers which were
screened and found to be effective at very low quantities.
In 1966, Harris, Chester, and Allen presented a very interesting review of more than 350 papers published during the
last twenty years on this subject.

The several polymers and related products, variously
proposed for soil conditioning, can be classified as

follows:
  a) Non-ionized polymers: polyvinylalcohol (PVA)
  b) Polyanions: polyvinylacetate (PVAc)
                  polyacrylonitrile, partly hydrolyzed
                  (H_pPAN)
                  polyacrylonitrile, hydrolyzed (HPAN)
                  polyacrylic acid (PAA)
                  vinyl acetate-maleic acid copolymer
                  (VAMA)
  c) Polycations: dimethylaminoethylmetacrylate (DAFMA)
  d) Strong dipolepolymers, inducing positive or negative
     bonds: polyacrylamide (PAM)
In this list, the most recent addition is polyacrylamide
(PAM). It is, in fact, the only product which can give a
non-soluble polymer in the soil and of which the price may
be economically feasible when applied in certain combinations. The other products are high in price and therefore
still prohibitive in application. PVA and PVAc are relatively cheap but are often washed out of the soil after a
quantity of 300 mm of water has passed through.
  On the world market the monomer of polyacrylamide (PAM)
is down to $0.70 per kg. The monomer as such cannot be
used as a soil conditioner because it is toxic to plants.
Studies by Schamp (1971) and others have shown that the
greater the polymer chain length the greater the soil conditioning efficiency. As a general rule with soil conditioners, the optimal molecular weight is around 10^6.
Therefore PAM should be brought into contact with the soil
as a large polymer. This is possible when the polymer is
applied in powder form. However, since the monomer costs
less than one third as much as the polymer, efforts have
been made in several countries to induce polymerization in
the soil. It is important that the PAM in the soil should
retain less than 1% monomer, for otherwise the material remains poisonous. Experience by the author and his coworkers has shown that no more than 50 g per m^2 are needed
to achieve good surface soil aggregation and stabilization.
In that case the depth of penetration is not more than 5-7
cm with a concentration of 0.05%. The action of this product is known from the laboratory studies of Greenland
(1965) to be due to hydrogen bonding between edge hydroxyl
groups of soil particles and polymer amide. The interaction between clay domain and polymer is thus:

$$
\begin{array}{ccc}
 & H & O \\
 & \| & \| \\
edge - OH - - - N - &C - R \\
 & \| & \\
 & H &
\end{array}
$$

Van der Waals attraction exists between edge or face of the clay particles and polymer, in addition to the electrostatic bonds between the amide group and the negative charge of the clay particle. With long-chain polymers hooked on clays at different spots, cross linkage occurs so that the bonds are strengthened. PAM also bonds to quartz particles by adhesion. Thus electrostatic adsorption and adhesion are both involved in this kind of soil conditioning.

Using PAM, the soil environment can be modified by inducing both aggregation and stabilization, making the soil relatively hydrophylic so its water absorption and infiltration capacity are increased. This product is very promising for use in sandy soils as no other economical hydrophylic product is available for the moment.

Emulsions of Polymers as Soil Conditioners

In recent years, attention has been drawn to the possibilities of using emulsions as soil conditioners (De Boodt, 1968, 1970) because they offer certain interesting possibilities not offered by solutions of polymers. The main advantage is that when an emulsion is properly applied to a soil, the micelle consisting of the polymer migrates to the points of contact between the soil particles. So when the soil is cultivated and brought to maximal porosity at the critical moisture content, the structure can be stabilized by spraying a dilute emulsion of polymers on it. Since bitumen is perhaps the cheapest product which can be easily emulsified, it is understandable that trials have been carried out with such an emulsion to determine its efficiency as a soil conditioner.

The action of this type of soil conditioner depends on the mode of migration of the active micelles, as well as on their size and stability. In attempting to control soil particles in order to get a solid link, three aspects are of importance:

a) The parameters determining the mobility of the micelles. (Migration of the micelles toward the meniscus is necessary to form a good link between soil

particles, as indicated in Figure 2).
b) The parameters determining the geometry of the solid
   link between the soil particles after the liquid of
   the emulsion is evaporated.
c) The properties of the chemical substances which make
   up the link (e.g., their cohesive strength and adhesion to the soil particles).

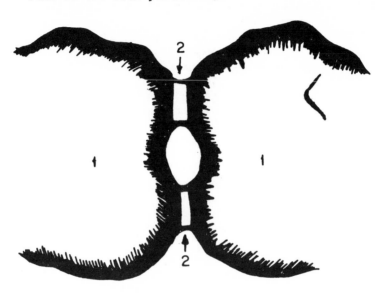

Figure 2. 1: sandy particle, 2: accumulation of the
micelles.

Detailed information on the physico-chemistry of soil
conditioning by emulsions was given by De Bisschop (1971).
An essential point is that in order to migrate the waterdrop in which the micelle is situated must hit a moist soil
so this waterdrop is no longer attracted to the soil particles but to the suction under the meniscus which is present
where two soil particles come in contact. With the waterdrop as the carrier, the micelles are thus drawn under the
meniscus. Through subsequent evaporation of soil moisture,
the emulsified micelles gradually become more concentrated
under the meniscus until the emulsion breaks, whereupon the
active material coagulates and links the adjacent soil particles together.
When it is desirable to make the soil more hydrophobic,

nearly unaltered bitumen can be used to form the micelles.
An emulsifier is used which imparts a small charge to the
colloidal particles and thus helps to keep them in suspen-
sion. If the bitumen is to be made hydrophylic, strong
acid groups must be attached to the micelles. This can be
carried out through sulfonation or with emulsifiers carry-
ing the sulfon group. Hydrophobic emulsions are particularly useful for stabilizing aggregation in soils susceptible to crust formation. In such soils germination of seeds
is a major problem, which can be obviated satisfactorily by
means of the soil conditioning treatment. This was proven
by experiments in Belgium on sugar beets and small grains,
as well as by trials carried out in Iran on irrigated
fields where sugar beets and cotton were grown.

Further benefits of this treatment result from the increase in saturated permeability (up to a factor of 5 or
10) and the consequent reduction of evaporation (as water-
films cannot form over the treated particle surfaces). It
can thus be said that a one-way screen for water is created
which contributes to the efficiency of water use. Most
striking results were obtained in the sandy desert of the
Gazvin region in Iran. On the plots treated with the hydrophobic emulsion, the moisture content under the treated
layer of 3-5 cm was twice as great as under the control
(untreated surface) and the natural growth of vegetation
was eight times as great as on the control. It bears
stressing that it was the treatment with hydrophobic emulsion which gave the best result and not the treatment with
hydrophylic emulsion, the results of which did not differ
significantly from those of the untreated plots, in terms
of moisture content and the growth of natural vegetation.

Application of Soil Conditioning Emulsion in the Field

Prior to application, the soil should be at its optimal
moisture content, i.e. the moisture content at which the
best aggregation of the soil can be obtained. This moisture content can be determined experimentally in the laboratory, or be estimated on the basis of farmers' experience.
Indeed, each farmer generally knows best at what moisture
content his particular soil can be crumbled and aggregated
most effectively. The sizes of the crumbs produced will
depend upon the types of soil and tillage implements.

Once it is properly structured, the soil is sprayed

with an emulsion the diffusion rate of which must be adjusted according to the type of soil and the composition of
the emulsion, in order to obtain the maximum penetration
and migration of the micelles to the bonding sites at the
interparticle contact points. The optimal dose varies between 0.5 and 1 liter of emulsion per square meter for use
in humid temperate zones and may be as high as 1 to 1.5 liter per square meter in arid zones. The main reason for
the higher dose generally required in arid conditions is
the high salt content and the higher osmotic pressure of
the soil solution at an increased moisture content. There
are also frequent differences in the nature of the clay between arid and humid-region soils.

The emulsions can be sprayed with ordinary spraying
equipment used for pesticides. However, the spray nozzles
should have a diameter of 0.8 to 1 mm and the pressure applied should reach 1.5 to 3 atm. With proper application,
the emulsion will not appear as a layer of paint on the
soil, but as a free liquid allowing the micelles to migrate
toward the contact points between the soil particles. To
enhance deep penetration, the emulsion-sprayed surface is
then mixed into the soil with harrow, to a depth of about
10 cm. In order to obtain sufficient structural stability
of the surface soil, it might be necessary in some cases to
pass over it with a cambridge-type roller. After 12 to 24
hours, the emulsion is completely broken and the micelles
have settled. The soil is then ready for sowing or planting.

Summary and Conclusions

From the foregoing it is evident that:
a) Only in recent years have adequate tools and means
   become available for studying the complex interactions among the different physical growth factors of
   the soil environment. From such studies one can
   hope to obtain information on which factors can and
   should be changed artificially.
b) Soil physical factors which are important to plants
   and are amenable to artificial modification are:
   water and air regimes, temperature, mechanical impedance, cation exchange capacity, and salt content.
c) In addition to classical soil treatments such as irrigation, drainage and tillage, new treatments have

been devised based on the use of chemical agents. These can allow us to obtain more stable aggregates, to make the soil either more hydrophylic or hydrophobic, to increase heat absorption, to enlarge the specific volume of the soil, and to either increase or reduce the cation exchange capacity and the salt content in the soil solution.
d) The artificial products which can produce these effects are either polymer solutions or polymer emulsions, each having its specific mode of action.

References

1. Bear, F.E. (1965). "Soils in Relation to Crop Growth," pp. 280-290. Reinhold, New York and Chapman & Hall, London.
2. Black, C.A. (1968). "Soil Plant Relationships," pp. 70-201. Wiley, New York.
3. De Bisschop, F. (1971). Physico-chemical aspects of soil conditioning with emulsions. Pedologie 21 (Gent), in press.
4. De Boodt, M. (1970). New possibilities for soil conditioning emulsions. FAO Meeting of European Commission on Agriculture, Working Party on Water Resources and Irrigation, 4th Session, Rehovot, 1-9.
5. De Boodt, M. and De Bisschop, F. (1970). Bodemstabilisatie en strukturatie met behulp van geëmulgeerde bitumen en polymeren. Cultuurtechnisch Tijdschrift 9 (Utrecht), 1-10.
6. De Boodt, M., Vandevelde, R., and De Bisschop, F. (1971). The slaking of the tilled soils, a major problem on mechanized farms. Rijksuniversiteit Gent.
7. Emerson, W.W. (1959). The structure of soil crumbs. J. Soil Sci. 10, 235-244.
8. Grable, A.R. (1966). Soil aeration and plant growth. Advan. Agron. 18, 57-101.
9. Greenland, D.J. (1965). Interaction between clays and organic compounds in soils, Part 1 and 2. Soils and Fertilizers 28, 415-425.
10. Harris, R.F., Chester, G., and Allen, O.N. (1966). Dynamics of soil aggregation. Advan. Agron. 18, 107-160.
11. Hénin, S. Feoforoff, A., Gras, R., and Monnier, G. (1960). Le Profil cultural. S.E.I.A., 129, Boulevard Saint Germain, Paris, 143-162.

12. Hillel, D. (1971). "Soil and Water: Physical Principles and Processes," pp. 201-224. Academic Press, New
York.
13. Kozlowski, T.T., ed. (1968). "Water Deficits and Plant
Growth," Vols. I and II. Academic Press, New York.
14. Netherlands Journal of Agricultural Science, Special
Issue (1967), 11 (Wageningen), 85-157.
15. Nielsen, B.F. (1963). Plant production, transpiration
ratio and nutrient ratios as influenced by interactions between water and nitrogen. Ph.D. Thesis, The
Roy. Veterinary and Agr. College, Copenhagen.
16. Problems of Soil Cultivation (1968). Proc. Int. Scientific Symposium, Brno, 1966.
17. Rawlins, S.L. (1971). Theory of high frequency irrigation: A new prospect for minimizing deep percolation.
Agron. J., in press.
18. Russell, E.J. (1950). "Soil Conditions and Plant
Growth." Longmans, Green and Co., London.
19. Salter, F.J. and Goode, J.E. (1967). Crop responses to
water at different stages of growth. Research Review
No. 2. Commonwealth Bur. of Horticulture and Planta-
tion Crops, Farnham Royal, Bucks, England.
20. Schamp, N. (1971). Soil conditioning by means of organic polymers. Pedologie 21 (Gent), in press.
21. Shaw, B.T., ed. (1952). "Soil Physical Conditions and
Plant Growth." Academic Press, New York.
22. Slayter, R.O. (1967). "Plant Water Relationships."
Academic Press, New York.

ROOT DEVELOPMENT IN RELATION TO SOIL PHYSICAL CONDITIONS

H.M. Taylor, M.G. Huck, and B. Klepper
Agricultural Research Service,
U.S. Department of Agriculture

Introduction

Plowing scenes are depicted on papyrus as early as the fifteenth century B.C. (Weir, 1936), and Virgil (70-19 B.C.) believed soil that was "...blackish and fat under the deep pressed share and whose mold is loose and crumbling is generally best..." (Tisdale and Nelson, 1966). Apparently people have long known that soil physical conditions can somehow affect crop production, but scientific studies of the significance of a well-developed root system were initiated only about three centuries ago.

Since that time, a large body of scientific literature has been produced on effects of soil conditions on root development. For example, Sutton's (1969) review of the form and development of conifer root systems contained about 30 pages of references. In this paper we do not attempt a review in the classical sense, because of the diversity and depth of research on the subject. Instead, we intend to review work conducted by the root-soil environmental relations group at Auburn, Alabama, and to relate that research to the work of others only when necessary to illustrate basic principles.

Our discussion concerns the effects of specific soil physical conditions on root extension. Experiments have been done both under controlled laboratory conditions where various soil factors have been studied separately, and under field conditions where factor interaction is important.

Root Growth

Increases in length of a particular root occur during primary growth when new cells that are formed in the apical region elongate and push the root tip through the surrounding medium. Turgor pressure in elongating cells must be sufficient to overcome the constraint of the cell walls and that of the external medium (Lockhart, 1965). The elongation process apparently is initiated in part by a "plasticization" of the cell wall, and stops when the cell wall again becomes rigid and develops secondary thickening. Thus, at least three factors are important in root elongation: turgor pressure within cells, the constraint offered by the cell wall, and the constraint offered by the surrounding medium. All three of these are affected by the soil physical environment surrounding the elongation region and the root tip. The extent and intensiveness of a root system can be determined by the restrictions imposed on root elongation rates by various soil horizons.

Branch roots develop from the pericycle in the root interior. They may either undergo secondary growth and remain as a permanent part of the root system, or die after a period of time. The lifetime of branch roots is probably controlled by endogenous factors, but is probably also modified by soil physical conditions. Control of secondary growth and factors affecting the formation and longevity of rootlets are all areas which need further clarification.

The pressure exerted by an elongating root against an external constraint has been termed "root growth pressure" by Gill and Bolt (1955). Root growth pressures developed by radicles of germinating seeds were measured during the past century when Pfeffer (1893) found axial growth pressures ranging from 12 to 25 bars for several species. In his experiments, the growth force tended to widen an air gap between two adjacent gypsum blocks containing an embedded root. Spring compression was used to measure the force necessary to maintain a uniform distance between the blocks. This air gap made data difficult to interpret because of the uncertainty of the effective cross-sectional area to be used in calculating pressure from force. With a similar air-gap technique Stolzy and Barley (1968) found that pea roots exerted growth pressures of about 6 bars. Taylor and Ratliff (1969a), still using a modification of Pfeffer's technique, but with many more replicates than previous investigators, measured average root growth pressures of 9.4, 13.0, and 11.5 bars for cotton (Gossypium hirsutum L.), pea (Pisum sativum L.), and peanut (Arachis hypogaea L.) seedlings, respectively. Like Pfeffer (1893), they concluded that the maximum root growth pressure was correlated with the osmotic potential of the root cells.

Techniques were improved when Eavis et al. (1969) devised a dead-load method which did not allow an air gap to develop, and thus gave data somewhat easier to interpret than those obtained previously. The average root growth pressure of cotton and peas was found to be about 11 and 12 bars, respectively. However, when gas composed of 3% O_2 and 97% N_2 surrounded the seedlings, the root growth pressure of cotton decreased to 5 bars. Thus, although radicles can exert axial pressures of at least 10 bars against external constraints, unfavorable soil conditions, such as low levels of O_2, can reduce these pressures. Unfortunately, no measurements have been made of growth pressures exerted by rootlets of mature plants.

Soil Temperature

Each species of plant has a minimum soil temperature below which no elongation occurs (Richards et al., 1952; Nielsen and Humphries, 1966; Trouse, 1971). Above that minimum temperature, root elongation rates increase almost linearly with temperature (Walker, 1969) to a maximum temperature which also depends upon the species. With further increases in temperature root elongation rates generally decrease rapidly. Nearly all of these conclusions were based on steady-state experiments where a specific soil temperature was maintained throughout the time period.

Since soil temperature varies diurnally under field conditions to depths of about one-half meter, root elongation probably also varies diurnally. It surely varies with major temperature changes caused by changes in weather and with depth in the soil profile. Ratliff and Taylor studied the time required for roots growing at a specific soil temperature to react to a change in temperature (Figure 1). Cotton was grown in root observation compartments (Pearson et al., 1970) filled with loamy sand soil at -1/5 bar water potential, 0.05 bars penetrometer resistance and negligible aluminum activity. All plants were grown at 32°C for 10 hours, then soil temperatures of 24°, 28°, 32°, and 36°C were imposed for an additional 70 hours. Root tip positions were marked every few hours. Time-lapse photographs were made at 3-minute intervals in accompanying studies. When soil temperature was altered, cotton taproot elongation responded to the new soil temperature within 15 minutes. The time lag in achieving a new equilibrium soil temperature precluded a closer determination of the lag period in root elongation rate. With the exception of the 36°C treatment, root length increased at the new rate almost linearly with time for 40 hours after temperatures were altered. For the first 10 hours after temperature was altered, roots at 36°C grew as fast as those at 32°C, but after 40 hours root length increases were at a slower rate at 36°C than at any other treatment temperature. The interaction of temperature level, root length increase, and time was also studied by Arndt (1945).

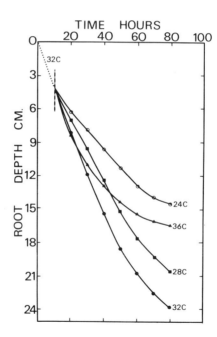

Figure 1.   Cotton root tip location during the first 70 hours after differential soil temperatures were applied. Each data point represents an arithmetic mean of 16 plants.

Most of the effects of changes in soil temperature on root elongation probably result from changes in root metabolic activity. However, effects of soil temperature changes in the viscosity of water, the hydraulic conductivity of the soil-root system, and cell-wall physics should not be ignored. Furthermore, many experiments in the current literature involve non-transpiring seedlings where the food supply is largely from stored materials. Controlled experiments are needed on the effects of temperature on elongation rates of rootlets of mature plants.

Soil Aeration

"Poor" aeration is known to affect both the form and the function of plant root systems (Russell, 1952; Van't Woudt and Hagan, 1957; Grable, 1966; Greenwood, 1969). However, considerable controversy exists about the aeration parameters that are most useful to measure, and about the levels of aeration that limit root development. Oxygen is consumed in respiration during root elongation; carbon dioxide and ethylene are two gaseous products of metabolic activity in roots. Through air-filled soil pores and channels within roots, oxygen is supplied and carbon dioxide and ethylene are removed. It is almost impossible, using current technology, to assess the relative importance of O_2, CO_2 and C_2H_4. Microtechniques are needed to measure partial pressures of these gases within root tissue.

The research group on root-soil environmental relations at Auburn, Alabama, has investigated three aspects of the aeration problem:

1. Tackett and Pearson (1964a,b) studied effects of O_2, CO_2 and soil bulk density on cotton root growth. Using O_2-N_2 mixtures, they found that root penetration through low density soil was not affected until O_2 concentration was reduced below 10%. Root penetration was decreased with reductions in O_2 concentration between 10 and 1.2%. In high density soil, O_2 concentration was unimportant in controlling root depth. Tacket and Pearson also investigated inhibitory effects of CO_2 on cotton root growth. At low soil bulk densities and 21% O_2, elongation rates decreased as CO_2 concentration increased from ambient air level to 24% CO_2. At high bulk densities, CO_2 concentration did not affect root elongation rates, probably because soil strength limited elongation.

2. Huck (1970) investigated effects of short-term fluctuations in soil oxygen status on radicle elongation rates of cotton and soybeans (Glycine max L. Merr.). He measured elongation rates while O_2 content of a gas stream that passed through the soil surrounding the roots was varied. Elongation ceased completely within 2 or 3 minutes after all O_2 was purged from the system (Figure 2) with 100% N_2 gas, and returned to normal shortly after 21% O_2 was reapplied to the system, if the period of complete anaerobiosis did not exceed 30 minutes. This rapid response to anaerobiosis implies that ATP or O_2 is directly involved in cell expansion, but further work is needed to determine the mechanism of this rapid cessation of growth. Periods of anaerobiosis longer than 30 minutes caused increasing proportions of taproot death until all were killed at 3 hours for cotton, and 5 hours for soybeans. This tissue death occurred within the region of elongation and did not extend to older tissue. Setting the nominal oxygen level at 3% resulted in an initial reduction in the rate of taproot extension (Figure 3), but elongation rates gradually returned to near those at 21% O_2. In his experiments, Huck found that lateral root initiation and development after a period of anaerobiosis occurred immediately above the killed portion. He concluded that distribution of roots through a volume of soil could be significantly influenced by anaerobiosis of even a few hours. This short-term condition might happen in flooded soils or perhaps in soils with perched water tables caused by the presence of tillage pans.

3. Eavis, et al. (1971) determined the effect of longitudinal internal diffusion of oxygen on pea radicle elongation. When air was maintained at 21% oxygen around the shoot, sufficient oxygen diffused through the plant pathways for radicles in humidified N_2 gas to elongate at a rate 20% of the control, i.e., where entire seedlings were maintained in humidified air. When the radicles were at21% 21% oxygen, but the cotyledons were in nitrogen, the radicle elongation rate was 50% of the control's. These data indicate that oxygen moves both up and down root tissue. Eavis[1] has obtained similar results for CO_2 and ethylene.

---

[1]Private communication.

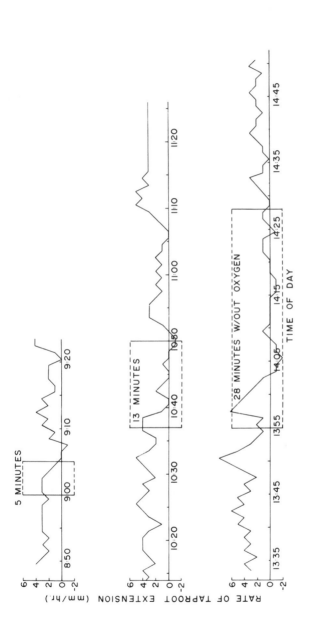

Figure 2. Elongation rate of a single taproot approximately 2 days after initiation of germination. Times shown are actual clock time on day of treatment, so the tracing represents successive periods of anaerobic treatment of the same root on the same day (Huck, 1970).

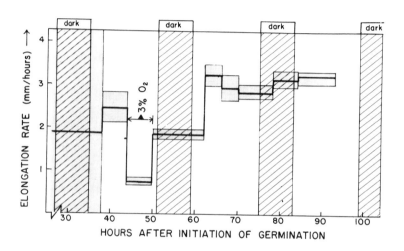

Figure 3. Elongation rate of a single cotton taproot as affected by passage through the growth medium of a gaseous mixture containing 3% O_2 and 97% N_2 (Huck, 1970).

Soil Water Status

Root growth is reduced as soil water potential is decreased (Klute and Peters, 1969). One might suppose that
the major factor involved is a reduction in hydraulic conductivity as soil water potential is decreased. However,
soil strength often increases with decreasing soil water
potential or water content, and most authors have not separated the effects of changes in soil water potential per se
from those of changes in soil strength.

Taylor and Ratliff (1969b) grew cotton in root observation boxes (Pearson et al., 1970) filled with loamy sand
soil at several water contents and several soil strengths,
which were estimated from penetrometer resistances. Taylor
and Ratliff (1969b) showed that root length of cotton grown
at 32°C and 0.05 bars penetrometer resistance was not affected by water potentials between –0.17 and –7.0 bars during the first 100 hours of elongation (Figure 4). At a

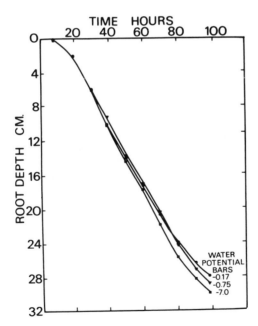

Figure 4. Cotton root tip location during the first
100 hours of growth at three soil water potentials. Each
data point represents the arithmetic mean of 16 plants.

given penetrometer resistance, root elongation rates were
the same at water potentials between –0.17 and -7.0 bars
for cotton (Figure 5) and between –0.19 and -12.5 bars for
peanuts. However, their experiments were conducted on germinating seedlings where transpiration was not allowed; experiments on roots of transpiring mature plants are needed.
In any event, many of the experiments which led to conclusions that root elongation rates were reduced with decreasing soil water potential should be reevaluated to see
whether the reduction resulted from the direct effect of
the soil water state or from the attendant increase of soil
strength.

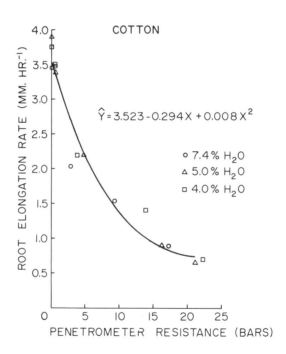

Figure 5. Effect of penetrometer resistance and soil
water content on cotton root elongation for the growth period 40 to 80 hours after radicle emergence (Taylor and
Ratliff, 1969b).

Many experiments have shown that plant roots will not extend into dry soil. However, definitive experiments have not been conducted to determine the rate of root extension as a function of water potential within mature plants or within the soil. In view of the fact that plant roots may shrink diurnally (Huck et al., 1970) experiments should also be conducted to evaluate soil-root transfer coefficients at various plant water potentials.

Soil Strength

Lutz (1952), Greacen et al. (1969), Eavis and Payne (1969), and Taylor and Bruce (1968) have reviewed effects of increased soil strength on root growth and development. In general, decreased root growth is associated with increased soil strength, but many unresolved questions remain.

Taylor and Ratliff (1969a) investigated the length of time required for cotton roots to change elongation rate in response to a change in constraint offered by the medium surrounding the root tip. They grew cotton in root observation compartments filled with a loamy fine sand soil at -1/5 bar water potential and 32°C. In each compartment, a 30 cm depth of soil was uniformly compacted, then a 2 cm layer of loose soil was added. When the tips encountered soil with penetrometer resistance of 5.8 and 11.5 bars, elongation rates immediately decreased (Figure 6). The decreased elongation rates associated with increased soil strength persisted for at least 100 hours. Information is needed on the changes in water potential in the elongation regions as the root tip encounters a constraining layer.

In other studies, time-lapse cinematography showed that the diameter of the meristematic and elongation regions increased when the tip encountered resistant material. After a possible lag period, the root tip then started moving downward at a rate which depended on the strength of the material. Histological sections (Camp and Lund, 1964) demonstrated that axially-stressed cells are not elongated but rather expand radially when roots grow in a compacted medium. However, histological research is needed to determine if roots stressed both axially and radially will react similarly to those stressed only axially.

The maximum root growth pressure available to force root tips through soil is less than 13 bars for cotton and

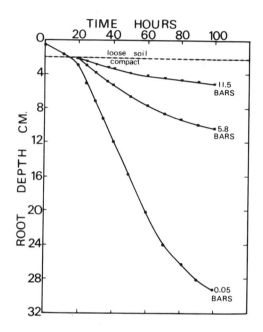

Figure 6. Cotton root tip location during the first 100 hours of elongation through soils that were compacted to three different penetrometer resistances. Each data point represents the arithmetic mean of 16 plants.

peanuts (Taylor and Ratliff, 1969a; Eavis et al., 1969). However, cotton and peanut root tips penetrated soils with a penetrometer resistance of 20 to 30 bars at about 20% of the maximum elongation rate (Taylor and Ratliff, 1969b). Reasons for the different reactions of the root and penetrometer tips are under active investigation at Auburn and other locations.

The experiments of Pearson et al. (1969) and the experiment illustrated in Figure 5 showed that seedling taproot elongation rates were very sensitive to increases in soil strength when other growth conditions were satisfactory. However, the question remained whether root development throughout a growing season would similarly react to the presence of a high-strength soil layer.

To answer this question, an experiment was conducted in the Auburn rhizotron (Taylor, 1969). Compartments 188 cm deep, 60 cm front-to-rear, and 120 cm side-to-side were filled with screened and fertilized surface soil with a pH of about 6.0. One compartment was filled with Fuquay loamy sand and one with Decatur clay loam. A third compartment was filled with Fuquay loamy sand until the soil depth was 165 cm. The surface was then thoroughly compacted by human foot traffic of about 0.25 bars pressure (77 kg man with footprints of 316 cm^2 area). After the foot traffic, a 23-cm depth of loose Fuquay soil was added as a seedbed. Penetrometer resistances, which were obtained with a 0.318-cm diameter moving point probe (Taylor and Ratliff, 1969b), were about 0.6 bars for all depths of the Decatur clay loam and the loose Fuquay loamy sand soils at -1/3 bar water potential. In the compartment where Fuquay loamy sand was compacted, penetrometer resistances at -1/3 bar water potential averaged 0.6, 5.5, 4.3, 2.1, and 1.2 bars at the 15, 30, 45, 60, and 180 cm depths, respectively.

Each of the compartments had four plants of cotton ('Auburn 56') equally spaced from side to side, but near the observation panel. Plant height, depth of deepest root and number of roots crossing horizontal transects at various depths were measured several times during the growing season. Plants in the Decatur clay loam grew slightly faster in height than those in the noncompacted Fuquay loamy sand (Figure 7). Forty days after planting, plants on the compacted loamy sand were significantly smaller than those in the other two compartments. An examination of the root system showed that no roots penetrated the compacted layer until 20 days after they had penetrated an equivalent depth in the other two compartments. Eighty days after planting, roots had reached the 188-cm depth in all three compartments.

Root development in the three compartments is shown graphically in Figure 8. Compaction significantly increased the number of roots visible at the 15-cm depth at all times greater than 20 days and less than 90 days (compare Figure 8a to 8b). Ninety days after planting, more roots occurred at depths greater than 75 cm in the clay loam soil (Figure 8c) than in the non-compacted loamy sand soil (Figure 8b). This increased rooting presumably occurred because of the greater quantities of available water during a low rainfall period in late July and early August,

1969. The gap in root data shown in Figure 8c occurred during a time-lapse photography sequence where Huck et al. (1970) showed that the cotton roots shrank diurnally.

The data collected in these three rhizotron compartments showed that a high-strength soil pan excluded root growth through the pan for about 20 days. This decreased depth of rooting was associated with a decreased plant height during most of the growing season. Plants on a clay loam soil grew taller than those on a loamy sand soil, presumably because of the greater quantity of available water in the clay loam soil during a period of low rainfall. Since these experiments were not replicated, effects of soil strength on yield could not be evaluated.

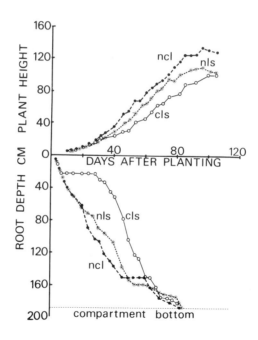

Figure 7. Height of the terminal point and location of the deepest root at various times after planting in noncompacted loamy sand (nls), noncompacted clay loam (ncl) and in loamy sand where a compacted layer existed at the 23-cm depth (cls).

70

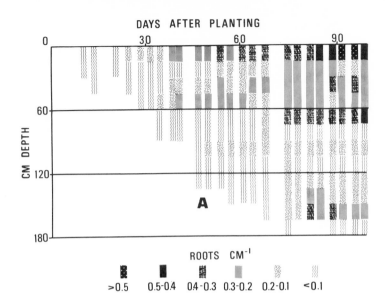

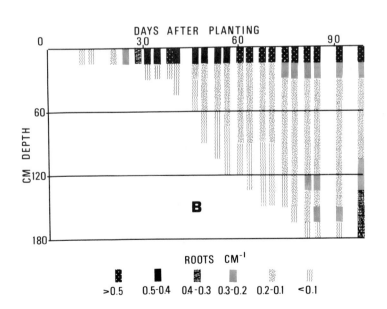

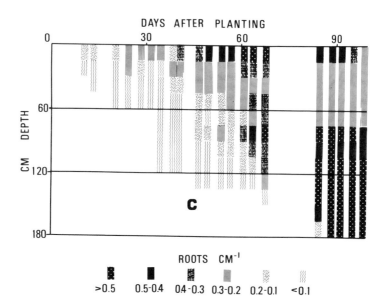

Figure 8. Cotton root concentration at various depths and times. Root concentration was measured by counting the roots that crossed 100-cm horizontal transects along the transparent plastic wall of a rhizotron compartment. The cotton was grown in (a) loose Fuquay loamy sand; (b) Fuquay loamy sand with a compacted layer starting at a 23-cm depth; and (c) loose Decatur clay loam.

Lowry et al. (1970) studied the effects of soil bulk density and depth to the soil pans on growth rate and yield of cotton in the field. They concluded that final plant height and yield of seed cotton were reduced as soil bulk density or penetrometer resistance increased. At equivalent soil bulk densities, cotton grown on pans located at the 10 cm depth produced less seed cotton than that grown on pans at 20 or 30 cm depths (Figure 9). A restricted water supply, due to limited rooting volume, was probably the principal factor causing reduced yield of cotton.

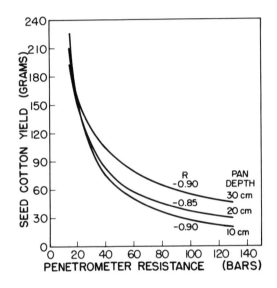

Figure 9. Seed cotton yield as a function of penetrometer resistance and depth to the soil pan layer. The "R" value on each curve is the correlation coefficient for the curvilinear regression equation of that depth (Lowry et al., 1970).

Interaction of Root Growth Factors

Relationships between root elongation rates of cotton seedlings and soil strength (Figure 5), soil temperature (Arndt, 1945), and soil solution aluminum activity (Adams and Lund, 1966) were developed in single-variable experiments. Pearson et al. (1970) investigated the interactions which might occur in taproot length 80 hours after planting when combinations of soil strength and aluminum activity, soil strength and soil temperature, and aluminum activity and soil temperature were varied systematically. Any of the three variables exerted the greatest effect when the other two were at optimum conditions, but when any variable was at or critically near a limiting value, the other variables exerted only a slight effect on elongation rate. There appeared to be no synergistic effects between any two of the variables.

Sixty-eight percent of the variation (R^2 = +0.68) in length of cotton taproots at 100 hours can be expressed by the equation

$$\hat{y} = -90.43T - 3.396P - 3.488 \log_{10}A + 3.186T^2 + 15.56 \text{ x}$$
$$10^{-2}P^2 - 36.93 \text{ x } 10^{-3}T^3 - 41.44 \text{ x } 10^{-4}P^3 + 36.50 \text{ x}$$
$$10^{-3}TP$$

where $\hat{y}$ = predicted root length (cm)
 T = soil temperature (C)
 P = penetrometer resistance (bars) with the Taylor-Ratliff (1969b) penetrometer
 A = aluminum activity in soil solution (μmolar) (Pearson, et al., 1970).

The equation illustrates the complexity of root growth, with eight coefficients needed to obtain a reasonable prediction of root length when only three of many possible root growth factors are studied.

Conclusions

Root elongation rates and root development are affected by a large number of factors. Experiments conducted by the root-soil environmental relations group at Auburn, Alabama, illustrate that root elongation rates respond rapidly to altered soil strength, soil temperature and soil aeration conditions. Long-term experiments have shown that increases in soil bulk density or soil strength will reduce the number of roots that penetrate soil pans and will slow the uptake of water and nutrients beneath pans. A reduced water supply often reduces top growth and yield.

An equation is presented (R^2 = 0.68) relating cotton root length at 100 hours to levels of soil penetrometer resistance, soil temperature and soil solution aluminum activity. The equation illustrates the complexity of root growth dependence on the soil environment. Many growth factors must be evaluated or controlled before root growth models can be effective tools in extrapolating knowledge.

References

1. Adams, F. and Lund, Z.F. (1966). Effect of chemical activity of soil solution aluminum on cotton root penetration of acid subsoils. Soil Sci. 101, 193-198.
2. Arndt, C.H. (1945). Temperature-growth relations of the roots and hypocotyls of cotton seedlings. Plant Physiol. 20, 200-220.
3. Camp, C.R. and Lund, Z.F. (1964). Effect of soil compaction on cotton roots. Crops and Soils 17, 13-14.
4. Eavis, B.W. and Payne, D. (1969). Soil physical conditions and root growth. In "Root Growth" (W.J. Whittington, ed.). Butterworths, London.
5. Eavis, B.W., Ratliff, L.F., and Taylor, H.M. (1969). Use of a dead-load technique to determine axial root growth pressure. Agron. J. 61, 640-643.
6. Eavis, B.W., Taylor, H.M., and Huck, M.G. (1971). Radicle elongation of pea seedlings as affected by oxygen concentration and gradients between shoot and root. Agron. J. 63 (in press).
7. Gill, W.R. and Bolt, G.H. (1955). Pfeffer's studies of the root growth pressure exerted by plants. Agron. J. 47, 166-168.
8. Grable, A.R. (1966). Soil aeration and plant growth. Advan. Agron. 18, 57-106.
9. Greacen, E.L., Barley, K.P., and Farrell, D.A. (1969). The mechanics of root growth in soils with particular reference to the implications for root distribution. In "Root Growth" (W.J. Whittington, ed.). Butterworths, London.
10. Greenwood, D.J. (1969). Effect of oxygen distribution in the soil on plant growth. In "Root Growth" (W.J. Whittington, ed.). Butterworths, London.
11. Huck, M.G. (1970). Variation in taproot elongation rate as influenced by composition of the soil air. Agron. J. 62, 815-818.
12. Huck, M.G., Klepper, B., and Taylor, H.M. (1970). Diurnal variations in root diameter. Plant Physiol. 45,

529-530.
13. Klute, A. and Peters, D.B. (1969). Water uptake and root growth. In "Root Growth" (W.J. Whittington, ed.). Butterworths, London.
14. Lockhart, J.A. (1965). Cell extension. In "Plant Biochemistry" (J. Bonner and J.E. Vanner, eds.). Academic Press, New York.
15. Lowry, F.E., Taylor, H.M., and Huck, M.G. (1970). Growth rate and yield of cotton as influenced by depth and bulk density of soil pans. Soil Sci. Soc. Amer. Proc. 34(2), 306-309.
16. Lutz, J.F. (1952). Mechanical impedance and plant growth. In "Soil Physical Conditions and Plant Growth" (B.T. Shaw, ed.). Academic Press, New York.
17. Nielsen, K.F. and Humphries, E.C. (1966). Effects of root temperature on plant growth. Soils and Fertilizers 29, 1-7.
18. Pearson, R.W., Ratliff, L.F., and Taylor, H.M. (1970). Effect of soil temperature, strength and pH on cotton seedling root elongation. Agron. J. 62, 243-246.
19. Pfeffer, W. (1893). Druck und Arbeitsleistung durch wachsende Pflanzen. Abhandlungen der Königlich Sächsischen Gesellschaft der Wissenschaften 33, 235-474.
20. Richards, S.J., Hagan, R.M., and McCalla, T.M. (1952). Soil temperature and plant growth. In "Soil Physical Conditions and Plant Growth" (B.T. Shaw, ed.). Academic Press, New York.
21. Russell, M.B. (1952). Soil aeration and plant growth. In "Soil Physical Conditions and Plant Growth" (B.T. Shaw, ed.). Academic Press, New York.
22. Stolzy, L.H. and Barley, K.P. (1968). Mechanical resistance encountered by roots entering compact soils. Soil Sci. 105, 297-301.
23. Sutton, R.F. (1969). Form and development of conifer root systems. Tech. Commun. No. 7, Commonwealth Agr. Bur., Farnham Royal Bucks, England.
24. Tackett, J.L. and Pearson, R.W. (1964a). Oxygen requirements of cotton seedling roots for penetration of compacted soil cores. Soil Sci. Soc. Amer. Proc. 28, 600-605.
25. Tackett, J.L. and Pearson, R.W. (1964b). Effect of carbon dioxide on cotton seedling root penetration of compacted soil cores. Soil Sci. Soc. Amer. Proc. 28, 741-743.

26. Taylor, H.M. (1969). The rhizotron at Auburn, Alabama—A plant root observation laboratory. Auburn Univ. Agr. Exp. Sta. Circ. 171, pp. 1-9.
27. Taylor, H.M. and Bruce, R.R. (1968). Effect of soil strength on root growth and crop yield in the southern United States. Trans. 9th Int. Congr. Soil Sci., Adelaide 1, 803-811.
28. Taylor, H.M. and Ratliff, L.F. (1969a). Root growth pressures of cotton, peas and peanuts. Agron. J. 61, 398-402.
29. Taylor, H.M. and Ratliff, L.F. (1969b). Root elongation rates of cotton and peanuts as a function of soil strength and soil water content. Soil Sci. 108, 113-119.
30. Tisdale, S.L. and Nelson, W.L. (1966). "Soil Fertility and Fertilizers." MacMillan Company, New York.
31. Trouse, A.C., Jr. (1971). Effects of soil temperature on plant activities. In "Compaction of Agricultural Soils" (W.M. Carleton, et al., eds.). Amer. Soc. Agr. Engs., St. Joseph, Michigan.
32. Van't Woudt, B.D. and Hagan, R.M. (1951). Crop responses at excessively high soil moisture levels. In "Drainage of Agricultural Lands" (J.N. Luthin, ed.). Amer. Soc. Agron., Madison, Wisconsin.
33. Walker, J.M. (1969). One degree increments in soil temperatures affect maize seedling behavior. Soil Sci. Soc. Amer. Proc. 33(5), 729-736.
34. Weir, W.W. (1936). "Soil Science, Its Principles and Practice." Lippincott, Chicago, Illinois.

THE FIELD WATER BALANCE AND WATER USE EFFICIENCY

D. Hillel
The Hebrew University of Jerusalem, Israel

The Field Water Balance

A. General

The water balance of a field is an itemized statement of all gains, losses, and changes of storage of water occurring in a given field within specified boundaries during a specified period of time. The task of monitoring and controlling the field water balance is vital to the efficient management of water and soil. Without knowledge of the water balance, one cannot very well evaluate possible methods designed to minimize loss and to maximize gain and utilization of water, which is so-often the limiting factor of crop production.

B. The System

Before we can itemize the field water balance, we must specify the boundaries of our system. From an agricultural or plant ecological point of view, it seems most pertinent to consider the physical realm within which the crop grows, i.e. from the top of the canopy to the bottom of the root zone. For a mature, "closed" crop of stable configuration, the boundaries are relatively easy to define. However, for an "open" crop that is laterally inhomogeneous and for a crop that is changing in extent and configuration, the geometric definition of the system can become quite difficult. To simplify our present discussion, we shall assume herein that our system is stable and homogeneous (at least laterally). The relationships we shall describe are valid, in principle, in systems of different configuration as well, provided their boundaries are geometrically definable.

Our hypothetical field is illustrated in Figure 1.

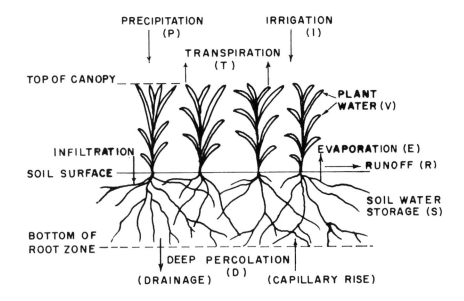

Figure 1. The field water balance (schematic),
(After Hillel, 1971).

C. Itemization of the Water Balance

Gains of water in the field are generally due to precipitation and irrigation. Occasionally, there may be
gains due to accumulation of runoff from higher tracts of
land, or to capillary rise from below (especially where a
water table is present at some shallow depth). Losses of
water include surface runoff from the field, deep percolation out of the root zone (drainage), evaporation from the
soil surface, and transpiration from the crop canopy. The
change in storage of water in the field can occur in the
soil, as well as in the plants. The total change in storage must equal the difference between the sum of all gains
and the sum of all losses.

Accordingly, we can state the water balance equation as
follows:

$$[\text{Gains}] - [\text{Losses}] = [\text{Change in Storage}]$$

$$(P + I) - (R + D + E + T) = (\Delta S + \Delta V) \tag{1}$$

wherein P is precipitation, I irrigation, R runoff from the field, D downward drainage out of the root zone, E evaporation from the soil, T transpiration by the crop canopy, ΔS the change in soil water content of the root zone, and ΔV the change in plant water content. All of these quantities are usually expressed in terms of water volume per unit of land area, i.e. in units of depth (cm).

D. Time-Period Considerations

An important consideration in water balance studies is the period, or time interval, for which the balance is made. Too short a period might be impractical, while too long a period might mask the occurrence of short-term critical stages. At such critical stages as flowering and fruit-set, even temporary imbalance in crop-water status (e.g., under the influence of a sudden hot-spell) can have a lasting effect. Suppose, for instance, that seasonal precipitation is sufficient to balance the climatic evaporative demand. However, since precipitation is episodic while the evaporative demand is practically relentless, a precipitation event might come too late to save the plants. It therefore may not avail us to have a positive balance for the whole season, and we must ensure that the crop water balance is maintained positive continuously during the growing season.

Evaluation of the Water Balance

A. General

In principle, a single equation can be solved for only one unknown, that is when all terms but one are given or can be measured. The solution of such an equation has a further limitation. If each of the measured variables entails an appreciable error, these errors can accrue to make the resulting solution grossly inaccurate. The relative error can become especially large in cases where a difference is sought between nearly equal quantities, each with an appreciable margin of error. Herein lies the traditional difficulty of employing the field water balance as a working tool in practical soil and water management. In many cases, the terms of the equation could not be measured with sufficient accuracy, or not at all. Fortunately, however, recent developments in soil physics and technology

81

now make it possible in many cases to monitor and evaluate
the field water balance with a sufficient degree of accuracy so that its use has become feasible in actual practice.

B. Special Cases

To be sure, there have always been some special cases
for which the water balance could be simplified and its
evaluation made easier by reducing the number of variables.
For example, if one considers a long period such as an entire year, then the change in water storage is likely to be
but a small, and perhaps negligible, fraction of the total
water balance, so that in this case the sum of precipitation and irrigation (P + I), assuming no runoff, is approximately equal to the sum of drainage and evapotranspiration
(D + E + T). Another case is a dry spell without precipitation or irrigation, during which the sum of drainage and
evapotranspiration must equal the change in storage.
Still other examples are the special cases in which
drainage is negligible, e.g. in the presence of an impervious barrier at the bottom of the root zone. In arid regions, it may happen that the scant precipitation wets only
the upper layer of the soil and is subsequently drawn up by
evapotranspiration without leaving any excess for drainage
beyond the reach of the roots, so that the root zone is un-
derlain by a "dry bottom" throughout the growing season.
In such cases, it is possible to assume that the diminution
of soil water content (ΔS) between rains or irrigations is
about equal to evapotranspiration (E + T), since ΔV is generally rather small relative to the other terms.

C. Measurement of Water Balance Components

In general, one cannot depend on the simplifications
given in the previous section, but must consider the water
balance in its entirety. This requires that each measurable item be evaluated independently and accurately. It is
relatively easy to measure the amount of water added to the
field by precipitation and irrigation, though it is necessary to consider non-uniformities in areal distribution.
Unfortunately, though the necessary instrumentation is generally available, field managers seldom take the trouble to
monitor the quantity and distribution of precipitation, and
particularly of irrigation, with sufficient accuracy.

The amount of runoff, in principle, should be minimal
in most agricultural fields, so that it is tempting to disregard this item. However, where runoff (or surface drainage, as it is often called) is appreciable, it is rather
difficult to measure and even more difficult to ascribe
quantitatively to different parts of the watershed above
the gauging point. Often, such a watershed consists of
more than one field.

Measurement of the water storage item, while not generally a problem theoretically, can be quite laborious and
expensive in practice. Storage of water in the crop is often disregarded, though not always justifiably, on the assumption that it is small in relation to water storage in
the soil. In any case, ΔV can be obtained by measurement
of growth and tissue hydration. Soil water storage and its
changes can be obtained by sampling, or by use of such instruments as electrical resistance blocks, or, preferably,
the neutron moisture meter. The great advantages of the
neutron meter over the traditional sampling method are
(Holmes et al., 1967) that it measures wetness on the volume basis directly, and that it samples a larger and hence
more representative mass of soil while minimizing sampling
errors (with repeated measurements made at the same sites)
and destructive augering. Moreover, the neutron meter provides an immediate answer, thus obviating the need for conveying samples to the laboratory to be weighed and oven-
dried, a procedure which is time consuming as well as tedious and a source of additional errors.

The three processes most difficult to measure directly
are evaporation, transpiration, and drainage. No distinction is generally made between the first two of these prc -
cesses, which are usually regarded as a single process and
termed evapotranspiration. Recently, methods have been
proposed to obtain separate measurements of the direct
evaporation of soil moisture and of transpiration of crop
canopies (e.g., Black et al., 1969). However, these twin
processes interact very strongly, and it seems doubtful
that they can be controlled independently of each other
(e.g., from heat budget considerations we might expect that
a reduction in soil moisture evaporation may result in a
nearly commensurate increase in transpiration). Though the
possibility of the separate control of evaporation and
transpiration deserves further study (see, for instance,
the chapter by Fuchs in this volume), it seems that we are

likely to have to continue, in the future as in the past, to work in terms of total evapotranspiration, however cumbersome the term may sound.

Of more immediate concern is the problem of obtaining separate determinations of evapotranspiration and of drainage. The water balance equation can be solved for the one only if the other is known (together with all additional terms of the equation).

D. Measurement of Evapotranspiration

Increasingly sophisticated techniques are now under development by micro-meteorologists to measure evapotranspiration. As reviewed recently by Tanner (1968) these include the energy balance-Bowen ratio method, as well as aerodynamic and eddy diffusion methods. So far, however, none of these methods has become practical as a working tool in field management.

The most direct method for measuring evapotranspiration is by use of lysimeters (e.g., Pruitt and Angus, 1960; Pelton, 1961; van Bavel and Myers, 1962; Black et al., 1968; Hillel et al., 1969). Such devices, when equipped with weighing and drainage mechanisms, can in fact measure the total water balance. However, to be accurate, lysimeters are necessarily bulky and rather expensive, and since they obstruct normal field operations they are not likely to be adopted in normal field practice. Even as research tools, lysimeters are problematic in that they seldom provide a reliable representation of the real above-ground and soil environments of the field in which they are set.

All of the available techniques notwithstanding, the direct measurement of evapotranspiration is still very difficult. If we wish to determine evapotranspiration from the water balance, we are left with the problem of measuring the drainage component of this balance.

E. Measurement of Drainage out of the Root Zone

In the past, measurement of deep percolation was such a formidable task that the temptation was great to shirk it completely by "assuming it away." A whole generation of soil physicists appears to have succumbed to this temptation and to have been guilty of negligence. The ideological justification for this was based on the concept of "field capacity," which held that downward drainage "virtually ceased" after two days or so, hence if one irrigated merely to fill the deficit to field capacity there would be essentially no net drainage. On the basis of this concept, numerous experimenters took to reporting as "evapotranspiration" the entire reduction of root-zone soil water content between irrigations in total disregard of the drainage component.

Recent studies (e.g., Gardner et al., 1970) have shown that the drainage process following an irrigation can persist for a very long time, albeit at a diminishing rate. Even if the water content of a given depth-layer (say, at the bottom of the root zone) remains constant, we cannot assume that water movement has ceased since the flow process may take place under nearly steady-state conditions. A recent study (Hillel and Guron, 1970) has shown that the deep percolation process can be bi-directional, i.e. alternately downward and upward in response to cyclic irrigations, and that the net drainage during an entire irrigation season can constitute 20% or more of the field water balance even under a "normal" water regime. Rather than to be assumed negligible, the drainage process must in fact be measured and controlled if we are to increase the efficiency of water management. Some drainage is, of course, essential in irrigated agriculture, particularly in arid regions, to prevent deleterious build-up of salts. Excessive drainage, on the other hand, involves unnecessary loss of water and leaching of nutrients, as well as of harmful salts. Determination of optimal drainage requirements in given conditions is a task for local research. General aspects of solute transport in the soil profile and of salinity management are discussed in the chapter by Bresler in this volume.

Direct measurement of the drainage component of the field water balance will become possible with the development of water flux meters (Cary, 1968). Such devices are still in the preliminary development stage, however.

A method is now available to determine the drainage component and thus allow computation of evapotranspiration from field measurements of soil moisture movement. This method is based on an initial measurement of the intrinsic hydraulic properties of a complete soil profile in situ. We shall now describe this method briefly.

F. Method of Determining Water Movement in the Soil Pro-
    file

The method we refer to has been termed the instantaneous profile method by Watson (1966). It requires frequent
(preferably continuous) and simultaneous measurements of
the soil wetness and matric suction profiles under conditions of drainage alone (evapotranspiration prevented).
From these measurements, it is possible to obtain instantaneous values of the potential gradients and fluxes operating within the profile, and hence also of hydraulic conductivity values. Once the hydraulic conductivity at each
elevation within the profile is known in relation to wetness, the data can be applied to the analysis of drainage
and evapotranspiration in a vegetated field. Work along
these lines has been carried out by Rose and Stern (1967a,
b), van Bavel et al. (1968a,b), Gardner (1970), and Giesel
et al. (1970).
The general equation describing the flow of water in a
vertical soil profile is

$$\frac{\partial \theta}{\partial t} = \frac{\partial}{\partial z} \; [K(\theta)\frac{\partial H}{\partial z}] \tag{2}$$

where θ is volumetric wetness (measurable by means of the
neutron meter), t time, z the vertical depth coordinate
here taken as positive downward, K the hydraulic conductivity which is a function of soil wetness, and H the hydraulic head (being the sum of gravitational and matric suction
heads, the latter measurable by means of tensiometers).
Integrating, we obtain

$$\int_{o}^{Z} \frac{\partial \theta}{\partial t} \; dz = \left(K \; \frac{\partial H}{\partial z} \right)_{Z} \tag{3}$$

or

$$\frac{\partial \theta}{\partial t} \; Z = \left(K \; \frac{\partial H}{\partial z} \right)_{Z} \tag{4}$$

Here Z is the soil depth to which the measurement applies.
If the soil surface is covered to prevent evaporation and
only internal drainage is allowed, the total water content
change per unit time (obtainable by integrating between
successive soil moisture profiles down to the depth Z) is
thus:

$$\left(\frac{dW}{dt}\right)_Z = \left(K\,\frac{\partial H}{\partial z}\right)_Z \tag{5}$$

Here W is the total water content of the profile to depth Z, i.e.

$$W = \int_o^Z \theta dz \tag{6}$$

Below the depth of the root zone, $(\partial H/\partial z)_Z$ (the hydraulic gradient at the depth Z) is often found to be unity; that is, the suction gradient is nil and only the gravitational gradient operates. If so, then $K = (dW/dt)_Z$. Otherwise, the suction gradient must be taken into account and the hydraulic conductivity is obtained from the ratio of flux to the total hydraulic head gradient (gravitational plus matric). This can be done successively at gradually diminishing water content during drainage, to obtain a series of K vs. θ values and thus establish the functional dependence of hydraulic conductivity upon soil wetness for each layer in the profile. The handling of the data and detailed method of computation have been described by Watson (1966).

To apply this method in the field, one must choose a characteristic[1] fallow plot that is large enough (say, 10 x 10 meters) so that processes at its center are unaffected by its boundaries. Within this plot, at least one neutron access tube is installed as deeply as possible to below the root zone. The required depth will usually exceed two meters. A series of tensiometers is installed near the access tube (far enough to avoid interfering with the neutron readings, yet near enough to monitor the same soil mass; a distance of 50 cm will usually do), at intervals not exceeding 30 cm to a depth as great as possible. Water is then ponded on the surface and the plot is irrigated long enough so that the entire profile becomes as wet as it can be. In the case of a uniform profile, this will mean

---

[1] To characterize the profile within the plot in relation to the field as a whole it is necessary to conduct a series of diagnostic tests, including texture with particular reference to clay content and type, soil moisture characteristic curve (θ vs. matric suction ψ), etc., according to horizons.

effective saturation. Tensiometer readings can indicate
when steady-state infiltration conditions have been
achieved. When the irrigation is deemed sufficient, the
plot is covered by a sheet of plastic so as to prevent any
water flux across the surface (evaporation or infiltration).
To minimize heat flux effect, the impervious surface can be
painted white or at least covered with a layer of loose
soil. As the internal drainage process proceeds, periodic
measurements are made of water content and tension throughout the profile. These readings must be taken frequently
at first (at least daily) but can be taken at greater time
intervals as the process slows down.

Once the relation of hydraulic conductivity to soil
wetness is known for the profile, it is possible to interpret water content and tensiometry data from a vegetated
field so as to compute the drainage component of the field
water balance. The actual rate of evapotranspiration (E_t)
can then be obtained from the relation:

$$\frac{dE_t}{dt} = \left(\frac{dW}{dt}\right)_{Zr} - \left(K\,\frac{\partial H}{\partial z}\right)_{Zr} \tag{7}$$

The first term on the right hand side of equation (7) is
the rate of total diminution of soil water content in the
root zone (to depth Zr) per unit time, and the second term
is the flow rate (flux) across the bottom of the root zone
(being the product of the hydraulic conductivity by the hydraulic gradient operating across the Zr plane. Since in
practice these computations will be made for finite time
periods, average rather than instantaneous values of gradient, flux, and conductivity must sometimes be used.

When the hydraulic gradient below the root zone is unity, which happens frequently but should not be taken for
granted, then equation (7) reduces to:

$$\frac{dE_t}{dt} = \left(\frac{dW}{dt}\right)_{Zr} - K_{Zr} \tag{8}$$

where K_{Zr} is the hydraulic conductivity prevailing at the
depth Zr a function of the wetness θ at that depth.

The limitations of the method described are in that it
is difficult to carry out and interpret where the soil is
heterogeneous, either vertically (i.e. layered) or

horizontally (in which case lateral components of flow may be appreciable). Moreover, the data of K vs. θ obtained from internal drainage are likely to encompass a rather limited range which may suffice for the interpretation of deep percolation but not of soil moisture evaporation processes. Despite these limitations, the method described appears to be practical in many cases in the field, and since it is based on in situ measurements of the actual profile, it is much preferable to all methods based on laboratory testing of the hydraulic properties of samples (even those designated as "undisturbed").

G. Field Studies

Rose and Stern (1967) studied the time rate of water withdrawal from different soil depth zones in relation to soil wetness and hydraulic properties taking into consideration the processes of drainage and plant-root uptake. They wrote the water balance equation for a given soil depth (assuming flow to be vertical only) for a given period of time (t_1 to t_2) in the following form:

$$\int_{t_1}^{t_2} (i - v_z - q_e)dt - \int_o^z \int_{t_1}^{t_2} (\frac{\partial \theta}{\partial t})dzdt = \int_o^z \int_{t_1}^{t_2} r_z dzdt \qquad (9)$$

where i is rate of water supply (precipitation or irrigation), q_e evaporation rate from the soil surface, v_z the vertical flux of water at depth z, θ volumetric soil wetness, and r_z the rate of decrease of soil wetness due to uptake by roots. The average rate of uptake by roots at the depth z is

$$\bar{r}_z = \frac{1}{t_2 - t_1} \int_{t_1}^{t_2} r_z dt \qquad (10)$$

The pattern of soil water extraction by a root system can be determined by repeated calculations based on the above equations for successive small intervals of time and depth. The total (cumulative) water uptake by the roots R_z is given by

$$R_z = \int_o^z r_z dz \qquad (11)$$

These relationships, which are analogous to those described in the previous section of this chapter, were used by Rose and Stern to describe the pattern of soil water extraction by a cotton crop in the field. The results indicated that nearly all water extraction by the crop took place from the top 30 cm during the early stages of growth and from the top 100 cm during the later stages.

A similar and detailed field study of water uptake by a crop was carried out by van Bavel et al. (1968a,b). The calculated root extraction rates agreed reasonably well with separate measurements of evapotranspiration obtained with lysimeters. Soil water movement within the root zone of a sorghum crop indicated initially a net downward outflow from the root zone but this movement later reversed itself to indicate a net upward inflow from the wet subsoil to the root zone above. Similar phenomena were observed under a corn crop by Hillel and Guron (1970).

Water Use Efficiency

A. Definitions

Any concept of efficiency must be based on a measure of the output obtainable from a given input. The efficiency with which water is used in irrigation can be defined and evaluated in different ways, depending on the point of view

For instance, one might define an economic irrigation efficiency as the financial return per amount of water applied or per amount of money invested in the water supply. Since economic realities vary widely from one country or location to another, this is hardly likely to become a universally meaningful index.

Another index of irrigation efficiency is that used by engineers, variously defined as the net amount of water "added to the root zone" or "used in evapotranspiration," as a fraction of the amount of water taken from some source (Hillel and Rawitz, 1971). Since this index of efficiency takes into account the losses of water incurred in conveyance from source to field (as well as in deep percolation) it is obvious that it will differ according to whether we are concerned with an entire project, an individual farm, or a particular field.

In the final analysis, the efficiency of irrigation is determined in the field, since it is here that the ultimate

aim of irrigation meets with success or failure. This topic has two aspects: (1) the technical aspect of applying
water with minimal waste, e.g. by avoidance of unnecessary
runoff and evaporation losses; and (2) the agronomic aspect
of water utilization by the crop. We shall concern ourselves herein only with the second aspect, which is generally described in terms of <u>water use efficiency</u>.

Even this concept, unfortunately, is often used in a
vague and ambiguous sense. Specifically, it can be defined
in at least two ways:

(1) <u>Crop water use efficiency (e_c)</u>: the ratio of crop
yield (Y) to the amount of water taken up by the
crop and used in growth (V) and transpiration (T).

$$e_c = \frac{Y}{V + T} \qquad (12)$$

(2) <u>Field water use efficiency (e_f)</u>: the ratio of crop
yield (Y) to the total amount of water used in the
field, including growth (V), transpiration (T), direct evaporation from the soil (E) and drainage
loss (D).

$$e_f = \frac{Y}{V + T + E + D} \qquad (13)$$

Crop yield itself can be defined in terms of total
growth (i.e. dry matter production) or in terms of the marketable product. Whether it be forage, grain, fruit, root,
or fiber, the marketable product often constitutes but a
fraction of the total dry matter produced.

Of the two indexes defined (crop vs. field water use
efficiency) the former is certainly of fundamental interest,
but the latter is of greater practical importance.

Possibilities of Control

Since e_c is a ratio, its value can be increased either
by increasing the numerator or by decreasing the denomina-
tor of the defining equation (12). The numerator, being
plant production, depends on such plant factors as gains
due to photosynthesis versus losses due to diseases and
pests. Hence water use efficiency can be influenced by
such means as pest and disease control, the choice of crop
and the genetic improvement (by selection and breeding) of
its productivity and adaptation to the particular environment; as well as by improvement of the water, air, and

nutrient supply to the roots, and of light and carbon dioxide supply to the foliage.

In particular, the supply of water in the field is much more easy to control than the demand. It must be noted, however, that generalization about plant growth in relation to water use is rather dangerous, since plant water uptake and growth are an integration of the combined influences of the many and fluctuating factors in the system, such as degree of vegetative cover and interception of light by foliage, wetness and exposure of the soil surface, availability of water at critical stages of growth, etc.

A. Control of Soil Moisture

As pointed out by Kramer (1969) one of the most conspicuous deficiencies in control of the plant environment is the inability to maintain the water supply at a uniform level. In irrigation practice, as well as in nature, water supply frequently fluctuates extremely. This problem was recognized long ago. Livingston (1908, 1918) attempted to remedy the situation by embedding hollow porous cones ("autoirrigators") in the soil and connecting them by means of tubes to a reservoir which automatically supplied water as it was removed. Another idea was the use of double-walled pots with space for water between the glazed outer shell and the porous inner wall through which water could flow into the soil (Wilson, 1929; Richards and Blood, 1934). Different researchers obtained conflicting results with these devices (Richards and Loomis, 1942; Read et al., 1962). Early attempts to control water supply in greenhouses were reviewed by Post and Seeley (1943). Among other methods described is the use of a subirrigator consisting of a perforated tube buried in the soil and connected to a constant-level reservoir, which could be set below the level of the soil so as to apply the water under suction.

B. Soil Moisture Availability

For many years, researchers in soil physics and plant physiology struggled with the goal of defining the upper and lower limits of soil moisture availability, i.e. "field capacity" and "wilting point." It has lately been shown (Gardner, 1966) that neither of these concepts is sufficiently precise and general for universal use. Water above and below these limits can be, and sometimes is, taken up by plants. Plant growth can be retarded long before the so-called permanent wilting point is reached, yet water extraction by plants may continue well beyond this point. Recognition of the inadequacy of these concepts comes from a better understanding of the energy relations and of the movement of water with the dynamic continuum of soil-plant-atmosphere.

Workers in the field have almost universally tended to accept, and many continue to follow, the original precepts of F.J. Veihmeyer and his associates in California regarding the virtually equal availability of soil water to crops within the entire range between "field capacity" and "permanent wilting percentage." The practical application of these precepts resulted in a regimen of infrequent irrigation which periodically caused the dessiccation of the surface zone of the soil with consequent effect on the root system. Practical limitations on the frequency of irrigation by the conventional methods and the vagaries of field conditions have made it difficult to examine these precepts critically in the field and to test alternative methods for the maintenance of high-moisture conditions in the root zone, particularly in the surface-layer of soil, by applying frequent or even continuous irrigation.

More recently, however, evidence has been accumulating (e.g., Viets, 1966; Rawitz, 1969; Hillel and Guron, 1970) that many crops show a very pronounced, non-linear increase of yield with added water if soil moisture suction can be kept very low, say in the range of 1/10 to 1/2 bar. This is provided, of course, that the soil is not kept so wet as to restrict aeration, nor is the irrigation quantity so excessive as to waste water and leach nutrients. Such conditions cannot easily be achieved and maintained with the traditional surface and sprinkler irrigation methods, and as a result these findings have only rarely been confirmed under field conditions.

C. Newer Irrigation Methods

Fortunately, newer irrigation methods have been developed and are being perfected, by which much-needed information can be obtained on the optimum soil moisture and irrigation regimes for various crops. We refer specifically to sub-irrigation under suction by means of porous tubes (Post and Seeley, 1943), and particularly to the technique of trickle (drip) irrigation which has lately come into vogue in Israel and some other countries. The latter technique consists of the slow delivery of water to the soil surface (or to some depth within the soil) from either a nozzle line consisting of a series of point-sources, or from a continuous line-source. Though known in principle for many years, this method has become economically feasible only with the recent development of low-cost, mass-produced plastic tubing and fittings. Since the delivery lines can be left permanently in the field and the discharge rates are low, the system makes it possible to apply small and well-controlled irrigation applications at will. The system is also adaptable to the simultaneous application of irrigation water and plant nutrients in solution, as well as to automation of initation and termination of irrigation. The high yields reported with trickle irrigation may indeed be due to the almost continual maintenance of a low-suction moisture regime in the root zone, which can effectively prevent the plants from ever being subjected to moisture stress.

Another interesting lesson to be learned from the apparent success of trickle irrigation is that it is not necessary to irrigate an entire field uniformly if all plant needs can be provided in a restricted volume of soil. With water applied from well-spaced point or line sources, the spatial distribution of water in the soil under trickle irrigation is often found to be highly non-uniform; nevertheless, plants tend to thrive in this regime, though their roots are trained to concentrate in clusters around the water sources.

D. Control of Water Supply

It now seems amply clear that subjecting plants to stress impairs production. Data cited by Slatyer (1967), for example, indicate that when plants are subjected to stress the stomata close down and assimilation is thus reduced. However, we still do not know enough about the functional dependence of plant production on moisture supply relative to potential evapotranspiration in the field. The discussions engendered by the postulations of de Wit (1958) regarding the linear relation of yield to transpiration are evidence of this fact. A significant study on the correlation of grain yield with water supply in the Great Plains of the U.S.A. was reported by Leggett (1959). He showed that a certain minimal, or threshold, supply of water is needed to produce any yield, and thus water use efficiency can rise rapidly once this supply is exceeded. It is this relation between grain yield and water supply that makes the storage of each additional centimeter of water in the soil so important in dryland farming. It has been the justification for the practice of fallowing land in alternating seasons, even though this practice is not very efficient in conserving precipitation (Viets, 1966). In an irrigated experiment with corn, Hillel and Guron (1970) obtained a doubling of water use efficiency when water supply was increased from about 70% to nearly 100% of potential evapotranspiration.

Traditionally, the great fallacy in water management has been the tendency to save water per unit of land area, in order to spread the available water as far as possible so as to "green up" more land. Some of the agricultural planners in Israel, as well as in other arid countries, have fallen prey to this fallacy.[2] We must remember that our basic aim is not to save water but to increase production efficiency by optimizing the water supply (and other environmental variables) so as to maximize plant response. The best chance of increasing production efficiency by water management is to obviate water stress and prevent water from becoming a limiting factor in plant growth. Once this is achieved, we can more effectively deal with some of the other variables which can constitute limiting factors.

---

[2]All too often, experimenters who attempted to compare crop response patterns to different quantities of irrigation, on finding no "significant" difference, concluded that the lesser irrigation quantity is as good as the greater quantity. Hence, they often recommended saving water by making the crop do with much less than potential evapotranspiration. This interpretation disregards the possibility that a lack of significant differences between treatments may be due to large experimental errors (heterogeneous field, imprecise measurements, etc.) which might prevent detection of real intrinsic differences. Moreover, failure to prove a statistical difference at, say, the 5% probability level, should not by itself be taken as proof of similarity between populations (or treatments).

E. Fertilization

One of the principal soil management practices affecting water use efficiency is the use of fertilizers. This subject was reviewed by Viets (1962). He pointed out that increased fertilization to correct nutrient deficiencies can result in large increases in yield while increasing the water use only slightly, if at all. This is particularly true when the soil moisture regime is kept in the wet (but not too wet) range with the water supply allowing the plants to meet the climatically-determined transpirational demand without water stress. On the other hand, under conditions of limited water, the addition of deficient nutrients, while beneficial in many and perhaps most cases, can sometimes result in reduced yields if faster water use causes the crop to run out of water at a critical stage of growth. Under the ·latter conditions, seasonal water use may be practically unaffected while water use efficiency decreases drastically. In general, it is to be expected that the most efficient utilization of both water and nutrients will be achieved if the soil moisture is maintained at a high potential without causing aeration or temperature problems.

F. Drainage Control

A major problem encountered in the effort to increase field water use efficiency is the difficulty of controlling the amount of downward drainage. The rate and amount of drainage are likely to increase as the root zone is maintained at a higher average moisture content. The problem is to balance the frequency and quantity of irrigation against the evapotranspirational demand and the requirement to avoid salinization. Recently, techniques have been developed to place a barrier layer of asphalt at some depth so as to intercept drainage and create a greater effective water storage capacity in the root zone. These techniques are discussed in the chapter by Erickson in this volume.

Need for Overall Control of the Plant Environment

Numerous environmental factors other than soil moisture content, potential, supply method, or fertilization can control the efficiency of water use, and hence their control and optimization is becoming increasingly necessary. We have already mentioned soil aeration. Another example is atmospheric composition (e.g., carbon dioxide enrichment has been found to promote assimilation and growth in certain greenhouse experiments).

Since, in general, evapotranspiration accounts for the greatest share of water use, it would seem that a greater effort should be directed toward finding means to reduce the evaporative demand. Control of climatic and other environmental factors above the ground is difficult in the open field, but perhaps not impossible. (Some aspects of this problem are discussed in the paper by Fuchs in this volume.)

Recent years have witnessed a dramatic increase in crop production in greenhouses and other types of enclosed spaces. This development can make possible the control of environmental variables to a degree heretofore unattainable. In particular, control of the temperature and humidity of the ambient air may result in the reduction of evapotranspiration. However, this will depend on controlling the energy balance of the entire system and particularly on alternative methods for dissipating the heat which will no longer be taken up in the form of the latent heat of evaporation.

References

1. Black, T.A., Thurtell, G.W., and Tanner, C.B. (1968). Hydraulic load-cell lysimeter, construction, calibration, and tests. Soil Sci. Soc. Amer. Proc. 32, 623-629.
2. Black, T.A., Gardner, W.R., and Thurtell, G.W. (1969). The prediction of evaporation, drainage and soil water storage for a bare soil. Soil Sci. Soc. Amer. Proc. 33 655-660.
3. Cary, J.W. (1968). An instrument for in situ measurement of soil moisture flow and suction. Soil Sci. Soc. Amer. Proc. 32, 3-5.
4. de Wit, C.T. (1958). Transpiration and plant yields. Versl. Landbouwk Onderz. No. 646, Wageningen, Netherlands.
5. Gardner, W.R. (1966). Soil water movement and root absorption. In "Plant Environment and Efficient Water Use" (W.H. Pierre et al., eds.), pp. 127-149. Amer.

Soc. Agron. and Soil Sci. Soc. Amer., Madison, Wisc.
6. Gardner, W.R. (1970). Field measurement of soil water diffusivity. Soil Sci. Soc. Amer. Proc. 34, 832.
7. Gardner, W.R., Hillel, D., and Benyamini, Y. (1970). Post-irrigation movement of soil water: I. Redistribution. Water Resources Res. 6, 1148-1153.
8. Giesel, W., Lorch, S., and Renger, M. (1970). Water-flow calculations by means of gamma-absorption and tensiometer field measurements in the unsaturated soil profile. In "Isotope Hydrology 1970." Int. Atomic Energy Agency, Vienna.
9. Hillel, D. (1971). "Soil and Water: Physical Principles and Processes." Academic Press, New York.
10. Hillel, D. and Guron, Y. (1970). The use of radiation techniques in water use efficiency studies. Res. Rept. submitted to Int. Atomic Energy Agency, The Hebrew Univ. of Jerusalem, Israel.
11. Hillel, D. and Rawitz, E. (1971). Soil water conservation. In "Water Deficits and Plant Growth," (T.T. Kozlowski, ed.), Vol. III. Academic Press, New York.
12. Hillel, D., Gairon, S., Falkenflug, V., and Rawitz, E. (1969). New design of a low-cost hydraulic lysimeter for field measurement of evapotranspiration. Israel J. Agr. Res. 19, 57-63.
13. Holmes, J.W., Taylor, S.A., and Richards, S.J. (1967). Measurement of soil water. In "Irrigation of Agricultural Lands," pp. 275-306. Agronomy 11, Amer. Soc. Agron., Madison, Wisc.
14. Kramer, P.J. (1969). "Plant and Soil Water Relationships: A Modern Synthesis." McGraw-Hill, New York.
15. Legget, G.E. (1959). Relationships between wheat yield, available moisture, and available nitrogen in eastern Washington dryland areas. Wash. Agr. Exp. Sta. Bul. 609.
16. Livingston, B.E. (1908). A method for controlling plant moisture. Plant World 11, 39-40.
17. Livingston, B.E. (1918). Porous clay cones for the auto-irrigation of potted plants. Plant World 21, 202-208.
18. Pelton, W.L. (1961). The use of lysimetric methods to measure evapotranspiration. Proc. Hydrol. Symp. 2, 106-134 (Queen's Printer, Ottawa, Canada, Cat. No. R32-361/2).
19. Post, K. and Seeley, J.G. (1943). Automatic watering

of greenhouse crops. Cornell Univ. Agr. Exp. Sta. Bul. 793.

20. Pruitt, W.O. and Angus, D.E. (1960). Large weighing lysimeter for measuring evapotranspiration. Trans. Amer. Soc. Agr. Eng. 3, 3-15, 18.

21. Rawitz, E. (1969). The dependence of growth rate and transpiration on plant and soil physical parameters under controlled conditions. Soil Sci. 110, 172-182.

22. Read, D.W.L., Fleck, S.V., and Pelton, W.L. (1962). Self-irrigating greenhouse pots. Agron. J. 54, 467-468.

23. Richards, L.A. and Blood, H.L. (1934). Some improvements in auto-irrigator apparatus. J. Agr. Res. 49, 115-121.

24. Richards, L.A. and Loomis, W.E. (1942). Limitations of auto-irrigators for controlling soil moisture under growing plants. Plant Physiol. 17, 223-235.

25. Rose, C.W. and Stern, W.R. (1967a). Determination of withdrawal of water from soil by crop roots as a function of depth and time. Aust. J. Soil Res. 5, 11-19.

26. Rose, C.W. and Stern, W.R. (1967b). The drainage component of the water balance equation. Aust. J. Soil Res. 3, 95-100.

27. Slatyer, R.O. (1967). "Plant Water Relationships." Academic Press, London.

28. Tanner, C.B. (1968). Evaporation of water from plants and soil. In "Water Deficits and Plant Growth" (T.T. Kozlowski, ed.), Vol. I. Academic Press, New York.

29. van Bavel, C.H.M. and Myers, L.E. (1962). An automatic weighing lysimeter. Agr. Eng. 43, 580-583.

30. van Bavel, C.H.M., Stirk, G.B., and Brust, K.J. (1968a). Hydraulic properties of a clay loam soil and the field measurement of water uptake by roots: I. Interpretation of water content and pressure profiles. Soil Sci. Soc. Amer. Proc. 32, 310-317.

31. van Bavel, C.H.M., Brust, K.J., and Stirk, G.B. (1968b). Hydraulic properties of a clay loam soil and the field measurement of water uptake by roots: II. The water balance of the root zone. Soil Sci. Soc. Amer. Proc. 23, 317-321.

32. Viets, F.G., Jr. (1962). Fertilizers and the efficient use of water. Advan. Agron. 14, 223-264.

33. Viets, F.G., Jr. (1966). Increasing water use efficiency by soil management. In "Plant Environment and Efficient Water Use" (W.H. Pierre et al., eds.),

D. HILLEL

pp. 259-274. Amer. Soc. Agron. and Soil Sci. Soc. Amer., Madison, Wisc.
34. Watson, K.K. (1966). An instantaneous profile method for determining the hydraulic conductivity of unsaturated porous materials. Water Resources Res. 2, 709-715.
35. Wilson, J.D. (1929). A double-walled pot for the auto-irrigation of plants. Bul. Torrey Bol. Club 56, 139-153.

CONTROL OF SOIL SALINITY

E. Bresler

Volcani Institute of Agricultural Research, Israel

Soil salinity is an important factor of the environment in which plants grow. In regions with arid and semi-arid climates the agricultural potential of irrigated land may be limited by accumulation of salts in the root zone. In these instances removal of salts from the soil is necessary before optimum crop production can be attained.

The formation of saline soils and the decline of early agricultural civilizations in arid and semi-arid regions are related mainly to improper management and salinity control (Reeve and Fireman, 1967). Bernstein and Hayward (1958) estimated that even in our century 20% of the irrigated land in the U.S.A. is affected by salinity. Moreover, highly productive soils can be readily transformed into nonproductive saline soils through a failure to control salinity, thereby increasing the vast area of land already affected by salinity. Control of salinity in soil is therefore essential to the operation of a permanently successful agriculture. It involves reclamation of salt-affected soils and maintenance or improvement of non-saline or reclaimed soils.

Whether salts accumulate or leach depends on the processes by which they move into, out of, or within the soil profile. Thus, the salt content in the root zone may be controlled by deliberately controlling the conditions at its boundaries which may regulate processes within this zone. The efficiency of control can be improved by application of existing quantitative theories and knowledge of the dynamic behavior of the soil-water-salt-plant-system. This chapter therefore places emphasis on a quantitative approach to treating the mutual relations between water and salt dynamics and the possibility of predicting salt behavior in non-sodic soils. Management practices for controlling soil salinity in such agricultural soils are also discussed.

Effect of Soil Salinity on Plants

The extent to which soil salinity has to be controlled is dictated by the response of agricultural plants to salinity. Numerous experimental data indicate that plants are strongly affected by the salt concentration of the soil solution within the root zone (e.g., Hayward, 1954; Wadleigh et al., 1951; Shalhevet et al., 1969; Bernstein, 1961 and 1965; Bierhuizen, 1969).

For most crop plants growth reduction is controlled mainly by the total salt concentration of the soil solution and is largely independent of the different salt constituents present in the solution (Reeve and Fireman, 1967; Wadleigh et al., 1951; Bernstein, 1961 and 1965; Richards, 1954). It is usually assumed that these growth suppressing effects are largely osmotic, caused by a decrease of the water availability to plants (Wadleigh and Ayers, 1945; Hayward, 1954) or by accumulation of salts in the plants due to its osmotic adjustment mechanism (Bernstein, 1961b). In addition to the general osmotic effects on plant growth some plants are especially sensitive to a specific toxicity such as the sensitivity of many fruit crops to sodium and chloride present in the soil solution (Bernstein, 1965). Information on the crop response function to any soil solution concentration is a prerequisite to the control of soil salinity.

Crop response to soil salinity has generally been expressed in terms of the electrical conductivity (EC) or a specific ion concentration of the saturation extract of the soil (Richards, 1954; Bernstein, 1961a, 1964, 1965; Shalhevet et al., 1969). The electrical conductivity has usually been taken as a measure of the osmotic pressure of the soil solution while the saturated extract is generally related to the soil water characteristic under field conditions, and is easily determined in the laboratory (Richards, 1954). For many soils the saturation percentage is twice the "field capacity" percentage. Knowledge of the salt concentration at "saturation" (as determined by the saturated paste method) and of the water retention properties of a soil make it possible to evaluate experimental data on plant response to salinity provided that the soil water regime as well as the other growth factors remain within their "optimal" range. Differences in soil water regime, stage of growth, fluctuations in the salt distribution with time and depth, soil fertility, climate and all the other growth conditions, should also be taken into account in evaluating data on crop response to any salinity control situation.

To include the effect of soil water regime fluctuations on plant response to salinity, Wadleigh and Ayers (1945) suggested that crops respond to the integrated total soil water potential, or suction (TS), which was defined as the sum of the osmotic and matric potentials of the soil solution. This concept is especially useful if one assumes that the salinity effect is primarily an osmotic effect which influences water uptake by plants by the same mechanism as does the matric potential. With this assumption in mind it is possible to consider the osmotic and matric potentials as additive in their effect on plant growth and consequently on crop yields.

From the salinity control point of view, the total potential concept may be more general than the EC_e salinity index. That is because salinity control may involve modification of the soil water regime which in turn will modify the plant response function to salinity. For example, more frequent irrigations cause the time average soil water content to be higher and thus the average actual salt concentration and water suction to be lower, without affecting the salt concentration as expressed in terms of the saturated paste extract. To take into account the interaction effect of soil salinity and soil water regime, Yaron et al. (1971) have suggested an estimation procedure of crop response function to soil water content and total salt concentration expressed in terms of total soil water suction (TS). It should be noted that under a given soil water regime the TS is conditioned by salinity, and the plant response to the actual salinity can adequately be described with reference to the solution concentration of the saturated extract. Application of this plant response in different water regime situations is possible as long as the actual solution concentration can be inferred with sufficient accuracy from the concentration of the saturated paste.

An additional consideration in the evaluation of the salinity-plant growth relationship is the fluctuation and variation pattern of salt concentration over time and space. Results of field and greenhouse experiments suggest that the average salinity of the main root-zone profile

103

represents all the effect of salinity on plants regardless
of the shape of the actual salinity profile (Shalhevet and
Bernstein, 1968; Shalhevet et al., 1969). Experimental data on the effect of salinity fluctuation with time on plant
response are very limited. Some published results showed
that the response was directly related to the number of
days various plants were under salinization regime (Shalhevet, 1970). For some crops it may be assumed that plant
response to salinity is related to the time-weighted average of soil salinity during short periods in the growing
season or throughout the whole growing period (Yaron et al.,
1971). Short period averages are especially important in
annual crops since some plants are more sensitive to salinity during germination and seedling stages than at later
growth stages (Richards, 1954).

Solute Movement in Soils

Soil salinity control depends largely on the ability to
predict space-time salt distribution. Macroscopic-scale
theories are generally satisfactory for prediction purposes.
In such theories the soil is considered to be a continuous
porous medium, whose representative elementary volume is
sufficiently large for its properties to be characterized
by statistical averages. It is usually assumed that diffusion and convective (or viscous) flow are the main mechanisms responsible for solute movement in soils.

A. Diffusion in Unsaturated Soils

Diffusion of a solute in a uniform medium of water is
described macroscopically by Fick's first law

$$J_o = -D_o \frac{dc}{dx} \tag{1}$$

where J_o is the amount (grams, moles, etc.) of solute diffusing across a unit area (cm^2) per unit time (sec), D_o is
the diffusion coefficient of the solute in water ($cm^2 sec^{-1}$),
c is the solute concentration ($g \cdot cm^{-3}$ of solution, or mole $\cdot$
cm^{-3}) and x is the distance (cm) along the direction of net
movement of the solute.

When a solute diffuses through water in the soil, the
electrical and porous properties of the soil must be considered. Several investigators (e.g., Porter et al., 1960;

Kemper et al., 1964; van Schaik and Kemper, 1966) have shown that the diffusion coefficient of a solute in the soil solution (D_p) is less than the diffusion coefficient in a free water system. The ratio between D_p to D_o depends on several geometrical, physical and chemical soil properties such as water content and viscosity, tortuosity of the diffusional path, and exclusion or adsorption of electrolytes. As soil water content decreases in a given soil, the cross sectional area available for diffusion becomes smaller, the path length increases, and the water viscosity as well as the negative adsorption becomes more important. Kemper and van Schaik (1966) have shown that the solute diffusion coefficient in a clay-water system is a positive exponential function of water content (θ) and for practical purposes is independent of the salt concentration. The functional relationship between D_p and θ is of a general type,

$$D_p(\theta) = D_o a e^{b\theta} \tag{2}$$

where θ is the volumetric water content, a and b are empirical constants. Olsen and Kemper (1968) stated that data collected on soils fit equation (2) reasonably well with b = 10 and "a" ranging from 0.005 to 0.001 depending on the surface area of the soil studied (sandy loam to clay). This simplified $D_p(\theta)$ relationship was found to be applicable in the range of water contents corresponding to the range of suction between 0.30 and 15 bars (Olsen and Kemper, 1968). This is the relevant range for most cases in which diffusional flow contributes to salinity control problems. Thus, the contribution of diffusion to macroscopic solute movement in unsaturated soils may be approximated by

$$J_d = -D_p(\theta)\frac{dc}{dx} \tag{3}$$

where D_p may be calculated by (2) and c is the solute concentration of the soil solutions.

B. Convection (Viscous) Solute Flow

If the solution is flowing through the soil, transport of the solute by convection depends on the velocity of flow. Because of the porous nature of the soil the actual pore water velocity is distributed about the average velocity in a manner depending on the distribution of pore sizes and shapes. It is generally assumed that the macroscopic transport of the solute by convection can be described by an equation that takes into account the two modes (or components) of transport: (a) the average flow velocity component and (b) the mechanical or hydrodynamic dispersion component. (The latter results from the variation in flow velocities among the soil pores, which causes the solute to be dispersed with respect to the average solution flow velocity.) This hydrodynamic dispersion effect is similar to diffusion in the sense that there is a net movement of the solute from zones of high to zones of low concentration. Most investigators assumed, therefore, that an equation similar to (1) is a good first-order approximation of the dispersion component in the convective solute flow, with the diffusion coefficient replaced by the hydrodynamic dispersion coefficient (D_h).

Many experimental and theoretical works (see Perkins and Johnston, 1963; Ogata, 1970) have shown that the magnitude of the mechanical dispersion coefficient (D_h) in a given porous material depends on the average flow velocity. For a simple non-aggregated porous medium (Passioura, 1971) D_h may be approximated by considering it as proportional to the first power of the average flow velocity,

$$D_h(\overline{V}) = D_m|\overline{V}| \qquad (4)$$

where $\overline{V}$ is the average interstitial flow velocity (cm sec^{-1}) and D_m (cm) is an experimental constant which depends on the characteristics of the porous medium.

The two modes of convective solute transport (that due to average flow velocity of the solution and that due to dispersion) have been described. Assuming that these two convective components are additive, the total amount of solute material transported across a unit area in the flow direction (x) is obtained by the sum of these two flow components:

$$J_h = -D_h(\overline{V})\frac{dc}{dx} + \overline{V}\theta c \qquad (5)$$

where J_h represents the total amount of solute per unit cross section of macroscopic area transported by convection in unit time (g or mole·cm^{-2}sec^{-1}). The first term represents solute flow due to dispersion and the second is solute flow due to average solution flow velocity.

C. Combined Diffusion and Convective Flows

The joint effects of diffusion and convective flow is described by a combination of equations (3) and (5) as follows:

$$J_s = -[D_h(\overline{V}) + D_p(\theta)]\frac{dc}{dx} + \overline{V}\theta c = -D(\overline{V},\theta)\frac{dc}{dx} + qc \qquad (6)$$

where D is the combined diffusion-dispersion coefficient (cm^2sec^{-1}) and q is the volumetric soil solution flux ($cm^3 cm^{-2}sec^{-1}$). Both terms in the right hand side of equation (6) are only an approximation owing to the values of the macroscopic quantities, D, V, θ and c being defined as spatial averages. Regardless of the approximate nature of (6), it has been useful in practice for predicting purposes.

In agricultural soils changes in water contents due to infiltration, redistribution, evaporation and transpiration of water cause water and salt to move simultaneously. Steady state flow conditions described by (6) occur very seldom in natural soil conditions. To establish the mathematical expression for the general transient conditions, equation (6) is combined with the conservation of mass principle, expressed by the continuity equation. This states that the rate of change of the amount of solute within a soil element must be equal to the difference between the amounts entering and leaving the element. Equating the difference between outflow and inflow to the amount of dissolved salt accumulated within the soil element for the more common unidirectional vertical flow leads to the expression

$$\frac{\partial(c\theta)}{\partial t} = \frac{\partial}{\partial z}\{[D_h(\overline{V}) + D_p(\theta)]\frac{dc}{dz}\} - \frac{\partial(\overline{V}\theta c)}{\partial z} \qquad (7)$$

Equation (7) is a mathematical statement of transient vertical diffusive-convective solute flow under the conditions described. It applies to solutes which do not interact with the soil and only when there is no loss or gain of salt other than by the flow described in (6). Solutions of (7) may be applicable to salinity control situations under fallow soil conditions when either the total salt concentration or a specific anion of interest is involved, provided that neither salt precipitation nor anion exclusion take place.

Nielsen and Biggar (1961) worked out a system with a steady state horizontal flow of water. Under these conditions $\overline{V}$ and θ are constants and equation (7) becomes

$$\frac{\partial c}{\partial t} = D\frac{\partial^2 c}{\partial x^2} - \overline{V}\frac{\partial c}{\partial x} \tag{8}$$

where $D = (D_h + D_p)/\theta$ is the dispersion coefficient obtained by fitting the breakthrough data to an analytical solution of (8) using the appropriate boundary conditions (Nielsen and Biggar, 1962). Biggar and Nielsen (1967) discussed the limitations of this analytical approach. They concluded on the basis of their data and boundary conditions, that for purposes of predictions "for some range of velocities, simple porous material and a reasonable magnitude of error some satisfaction may be obtained" by the method described.

In some salinity control situations a specific ion concentration in the soil solution is of interest. In such cases the ion may interact with the soil material. The most important interaction is the adsorption of cations to the clay fraction of the soil. In some extreme cases anion exclusion may also be important. When equilibrium between the ions on the exchange phase and in the soil solution is reached very fast, the amount adsorbed, or excluded, may be assumed to depend on the equilibrium solution concentration only. Introducing these assumptions into equation (7), applying the chain rule and rearranging gives:

$$[\theta(z,t) + b(c)]\frac{\partial c}{\partial t} = \frac{\partial}{\partial z}[D(\theta,\overline{V})\frac{\partial c}{\partial z}] - q(z,t)\frac{\partial c}{\partial z} \tag{9}$$

where $b(c)$ is the slope of the adsorption isotherm as a function of concentration, $q = \overline{V}\theta$ is the volumetric water flux, and D is the combined dispersion coefficient. Note that equations (9) and (7) are identical except for the addition of $b(c)$. The left hand side and the last term in the right hand side of equation (7) were expanded and the conservation of mass statement was used in changing (7) into (9). Note also that the function $b(c)$ is given in terms of volumetric water content and that $b(c)$ is positive for adsorption and negative for exclusion.

Effect of Soil Salinity on Water Flow

The dominant effects of macroscopic water flow on solute movement in soils have been emphasized. Macroscopic water flow in unsaturated soil under isothermal and isosalinity conditions is described quite well by Darcy's equation, adapted to unsaturated soil,

$$q = \overline{V}\theta = -K(\theta)\frac{dH}{dx} \tag{10}$$

where q is the volumetric water flux (cm^3 $H_2O \cdot cm^{-2} sec^{-1}$), $K(\theta)$ is the hydraulic conductivity of the soil (a function depending on θ alone) and H is the hydraulic head which is the sum of the suction head (h) and gravity head. The nonlinear partial differential equation (11) which describes unidirectional vertical water flow in unsaturated soil is obtained by combining the mass conservation principle with Darcy's law:

$$\frac{\partial\theta}{\partial t} = \frac{\partial}{\partial z}[K(\theta)\frac{\partial H}{\partial z}] \tag{11}$$

In this Darcy type water flow equation it is assumed that the hydraulic gradient is the only driving force which causes water to flow. However, the dynamic changes of salt concentration, due to mass movement of salt and water content fluctuations, may create an additional driving force due to osmotic gradients. Also, variation in salt concentration and composition may affect the hydraulic conductivity function, $K(\theta)$, due to density and electrical changes. Thus, in applying equations (7) and (11) to a given salinity control situation the mutual salt-water flow effects must be considered.

A. Effect of Salt Concentration Gradients

Kemper and Evans (1963) have shown that when a solute is completely restricted by the soil, water flow in response to an osmotic pressure gradient is the same as when a hydraulic pressure gradient is applied provided that they are of equal magnitude. For this case of complete restriction, the hydraulic conductivity coefficient K can be used for both hydraulic and osmotic potential gradients. Thus, the total volumetric water flux may be described by:

$$q = -K[\frac{dH}{dx} + \frac{1}{\rho g} \frac{d\pi}{dx}] \qquad (12)$$

where ρ is the water density, g is the gravitation constant and π is the osmotic pressure (which may be related to salt concentration by $\pi \simeq \phi CRT$ where ϕ is the osmotic coefficient of the solute). When the salt is only partially restricted by the soil equation (12) becomes

$$q = -K[\frac{dH}{dx} + \frac{\sigma}{\rho g} \frac{d\pi}{dx}] \qquad (13)$$

where σ is the osmotic efficiency coefficient, which ranges between 0 to 1 and may be interpreted as the degree of semipermeability of the soil. The value of σ will be 0 when salt concentration gradients will cause no water to flow and will be 1 for complete solute restriction when the osmotic gradients are as effective in causing water to flow as the potentially equivalent hydraulic gradients. The greater the restriction of the solute relative to the solvent, the greater will be the value of σ (Kemper and Evans, 1963).

Letey (1968) reviewed the limited experimental information on water movement in response to salt concentration gradients in unsaturated soils. He concluded that at low suction σ is very small and water flow due to osmotic gradients is negligible. At higher suctions the total amount of water flowing in response to salt concentration gradients is still very low but becomes large relative to the flow due to the pressure gradients. Letey (1968) suggested that an approximate value of σ at soil water suctions between 0.25 to 1 bar is about 0.03, whereas at suctions less than 0.25 σ can be assumed to be zero. No data are available at higher suctions. It seems therefore that under most salinity control conditions salt concentration effects on macroscopic water flow can be neglected in practice. This implies that transient water flow under these conditions is adequately described by equation (11).

B. Effect of Soil Solution Concentration and Composition on the Soil Hydraulic Conductivity

Swelling of soils containing clay and dispersion of the soil colloidal material alter the geometry of the soil pores and thus affect the intrinsic permeability of the soil. It may be deduced from the double-layer theory that

both swelling and particle dispersion increase as soil solution concentration and Ca/Na ratio decrease. The hydraulic conductivity is affected not only by the intrinsic permeability, but also by the properties of the soil solution,
such as fluid density and viscosity, which are also affected by the composition and concentration of solutes.
Investigations have confirmed that the hydraulic conductivity behaves accordingly, i.e. higher in concentrated solutions or high Ca/Na ratios and lower in dilute solutions or
low Ca/Na ratio.

Gardner et al. (1959) have reported that in low sodium
soil, the weighted mean soil water diffusivity was no more
than doubled as the solution concentration was increased
from 2 to 100 meq/liter, whereas in high sodium soils dif-
fusivities changed by a few orders of magnitude at similar
changes of salt concentration. Data obtained by Quirk and
Schofield (1955) showed similar results for ranges of concentration commonly encountered in soils. Effect of solution concentration on soil permeability in soil with divalent cations was negligibly small compared to monovalent
soil systems.

The U.S. Salinity Laboratory Staff (1954) have proposed
the term Sodium Adsorption Ratio (SAR), which is an approximate expression for the relative activity of Na ions in
exchange reactions in soils. The soil solution cationic
composition in SAR terms is commonly used to describe
the cationic composition effect on soil hydraulic conductivity. It has been shown by many investigations (e.g.
Quirk, 1957; Naghshine-Pour et al., 1970) that the hydraulic conductivity increases as the SAR, or the associated
Exchangeable Sodium Percentage (ESP), decreases and the solution concentration increases. This was found to be true
as long as SAR had a value of at least 10. For lower values of SAR the effect of electrolyte concentration on soil
hydraulic conductivity was negligibly small. Thus, as a
first order approximation in estimating water flow for salinity control of nonalkaline soils the effect of solution
concentration on soil hydraulic conductivity may usually be
neglected.

Because of the fact that high electrolyte concentration
affects soil permeability in sodic soils, it has been suggested that reclamation of sodic soils should start with
high salt concentration water to increase permeability and
facilitate leaching (Reeve and Bower, 1960; Reeve and

Doering, 1966).

Estimation of Salt Distribution in the Soil Profile Under Fallow Conditions

Estimation procedures which will be described here may be applicable to most salinity control situations under fallow field conditions or when the effect of vegetation may be neglected. The description of the procedures is restricted to the estimation of total salt concentration when neither salt precipitation nor uptake by plants take place. It may also be useful to estimate a specific ion distribution when its exclusion or adsorption may be neglected. This is a good approximation for anions such as chlorides or sulfates, which are commonly present in the soil solution.

The soil under consideration is nonalkaline. It is horizontally homogeneous with respect to water content and salt concentration. The effect of salt concentration on water flow is considered to be negligible ($\sigma = 0$, SAR < 10) and temperature fluctuations are neglected (isothermal conditions are assumed).

Under these conditions, the differential equation (11) governs vertical water flow in the soil system. Equation (11), subject to appropriate boundary conditions, can be solved by numerical methods with the aid of a computer. Solutions have been given for many specific processes of infiltration, redistribution, drainage and evaporation (Freeze, 1969). Hanks et al. (1969) described a general numerical method for estimating one-dimensional vertical water flow as it occurs under fallow field conditions. Solutions for water content, water suction, and volumetric water flux at any time and depth were obtained for the processes of infiltration, redistribution, drainage and evaporation. The method provided for hysteresis in the water content-suction head relationship to account for wetting and drying which may occur simultaneously in different parts of the profile. Data of the main branches of the hysteresis loop are therefore required, in addition to the $K(\theta)$ function, for the computation. Hanks et al. (1969) using this procedure, obtained close agreement between computed and measured values of wetness, suction, and flux.

The type of flow (either infiltration, redistribution or evaporation) is determined by the water flow conditions at the soil surface boundary. The boundary conditions that must be satisfied at the soil surface ($z = 0$) according to this method are:

$$q(0,t) = -K(\theta)(\frac{\partial h}{\partial z} - 1)\Big|_{z=0} \leq R(t) \tag{14a}$$

$$R(t) > 0, \ h(0,t) \geq 0, \ \text{or} \ \theta(0,t) \leq \theta_s \tag{14b}$$
for any t during infiltration;

$$R(t) = 0 \quad \text{for any t during redistribution;} \tag{14c}$$

$$R(t) < 0, \ h(0,t) \leq h_d, \ \text{or} \ \theta(0,t) \geq \theta_d \tag{14d}$$
for any t during evaporation.

where $q(0,t)$ is the volumetric water flux at the soil surface ($cm^3 cm^{-2} sec^{-1}$); h is the water suction head (cm); h_d and θ_d are the suction and water content, respectively, corresponding to air dry soil under given evaporation conditions; θ_s is the soil water content at saturation; and $R(t)$ is a given "potential" flux at the soil surface ($cm^3 cm^{-2} sec^{-1}$). The actual value of $R(t)$ is determined by the irrigation or rain intensity during the infiltration stage and by the free water evaporation rate during the evaporation stage.

The lower boundary at a depth Z is arbitrary, and must always be chosen such that it is below the root zone where the suction gradient approaches zero, $\partial h/\partial z\big|_{z=Z} = 0$. In a simple drainage case a zero suction ($h = 0$) at the water table depth could be a bottom boundary condition. To the surface and bottom boundary conditions the known initial conditions $h(z,0) = h_n(z,0)$ or $\theta(z,0) = \theta_n(z)$ must be added.

From the numerical solution of (11) together with the associated boundary conditions, values of water content and flux as a function of time and depths [$\theta(z,t)$ and $q(z,t)$, respectively], can be obtained. These values were used by Bresler and Hanks (1969) and Bresler (1971) to estimate salt flow simultaneously with water in unsaturated fallow soils. For this purpose equation (7) (Bresler, 1971) or a simplified form of it (Bresler and Hanks, 1969) was solved numerically by a finite difference method.

In order to solve (7) the values of $D_h(z,t)$ and $D_p(z,t)$ must be evaluated. With $\theta(z,t)$ already known from the solution of (11), the value of the diffusion coefficient $D_p(z,t)$, may be calculated from (2) with the values of a and b of Olsen and Kemper (1968). Similarly, since $q(z,t)$

113

have been already calculated $\bar{V}(z,t) = q(z,t)/\theta(z,t)$ is used to evaluate the hydrodynamic dispersion coefficient (D_h) from the $D_h(\bar{V})$ relationships (4). The coefficient D_m in (4) may be approximated by a best fit breakthrough curve to the solution of (8), when $\bar{V}$ and θ are constant (Passioura et al., 1970), or from theoretical considerations (Perkins and Johnston, 1963).

Using the above-mentioned approximations, equation (7), subject to suitable initial and boundary conditions, was solved numerically with the aid of computer techniques (Bresler, 1971). The initial and boundary conditions which may be considered appropriate to a salinity control situation are as follows:

$$c(z,0) = c_n(z) \tag{15a}$$

$$-\{D_p[\theta(0,t)] + D_h[\bar{V}(0,t)]\}\frac{\partial c}{\partial z} + q(0,t)c(0,t)\} = q(0,t)C_o(t)$$
at $z = 0$, for any $t > 0$ during infiltration $\tag{15b}$

$$\left.\frac{\partial c}{\partial z}\right|_{z=0} = 0 \quad \text{for any } t > 0 \text{ during redistribution} \tag{15c}$$

$$-\{D_p[\theta(0,t)] + D_h[\bar{V}(0,t)]\}\frac{\partial c}{\partial z} + q(0,t)\}_{z=0} = 0 \tag{15d}$$

for any $t > 0$ during evaporation

$$\left.\frac{\partial c}{\partial z}\right|_{z=Z} = 0 \quad \text{for any } t > 0 \tag{15e}$$

where C_o is the salt concentration of the water applied, $c_n(z)$ the predetermined initial salt concentration profile, and all other symbols as defined before. It should be noted that the volumetric water flux at the soil surface $[q(0,t)]$, is negative during the evaporation stage.

To obtain salt distribution estimates $c(z,t)$ by the numerical method described, the $D_p(\theta)$ and $D_h(\bar{V})$ functions must be correctly known or approximately estimated. However, for many practical salinity control purposes it may be assumed that under transient conditions (7) the overall diffusion-dispersion term (see [6]) contributes very little compared to the macroscopic-average viscous flow (i.e., $D(\bar{V},\theta)dc/dz \ll qc$). Applying this assumption to (7) and (15) Bresler and Hanks (1969) solved the expression

$$\frac{\partial}{\partial t} \int_0^z [c(z,t)\theta(z,t)]dz = q(0,t)C_o(t) - q(z,t)c(z,t) \qquad (16)$$

Equation (16) was solved explicitly for $c(z,t)$ by a numerical method where values for $\theta(z,t)$ and $q(z,t)$ were obtained, as before, from the solution of (11) when the surface boundary conditions and the initial conditions are known.

Figure 1 demonstrates the results obtained from the computation of (7) and (15) by Bresler (1971), as compared to data calculated from (16) according to Bresler and Hanks (1969). The soil hydraulic characteristics used were those of Gilat loam soil (Bresler et al, 1971). The parameter data for the computation of (7) and (15) were: $D_m = 0.1$ cm, $D_o = 1.1 \times 10^{-5}$ cm/sec, a = 0.0025 and b = 10. The values of θ ranged between 0.03 in air-dry soil to 0.44 at saturation with the highest average flow velocity $(\bar{V})$ of 6×10^{-3} cm sec^{-1}. The comparison (Figure 1) suggests that, at least for the conditions studied, the additional approximation made by Bresler and Hanks (1969) can be safely used for estimation of salt distribution under fallow conditions.

Estimation of Salt Distribution in the Soil Profile Under Crop Growing Conditions

The assumption that salt uptake by plants might be neglected could also be applied in the presence of plants. However, water uptake by plants is a dominant factor and must be considered when salt distribution under crop growing conditions is of interest. Under these conditions, a macroscopic water extraction term must be added to equation (11) to account for water loss due to transpiration.

Molz and Remson (1970) suggested an extraction term depending on water content, soil depth and transpiration rate. The diffusion type of equation (11), including this term, was solved and compared with experimental data. The comparison was quite acceptable. Since the theoretical formulation of water extraction by plants is only at the beginning stages of development empirical models are useful. Yaron et al. (1971) found that the empirical relation S = $k_1(z,t) + k_2(z,t)\theta$ represents quite well the water extraction (S) throughout the root zone during the growing season of wheat in Gilat, Israel. The parameters k_1 and k_2 were estimated by a simulation model for discrete soil layers, crop growth stages and climatic conditions. Introducing

115

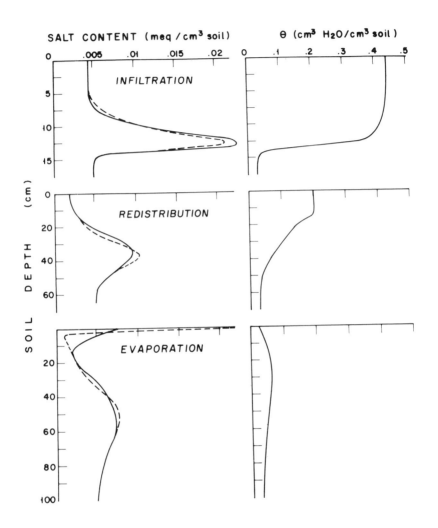

Figure 1. Computed salt and water content distribution. The dispersion term was considered (solid lines) and omitted (dashed lines). See text for a full explanation.

either the empirical or theoretical extraction term into
(11) and solving it numerically, it is possible to estimate
the salt distribution as before provided that the required
water extraction parameters are available.

The degree of accuracy that one obtains with the described methods in estimating the soil salinity variables,
$c(z,t)$, $\theta(z,t)$ and $h(z,t)$, depends primarily on the following factors: (a) proper choice of the mathematical model
to give an adequate description of the physical system at
hand; (b) proper specification of boundary conditions;
(c) accuracy and stability of the numerical procedure; and
(d) accuracy of the soil and plant parameters used in the
computation. Of these factors, the most critical one is
the accuracy of the soil and plant parameters.

Because of the difficulties involved in obtaining all
the necessary soil-water-plant-salt parameters required to
estimate the salt distribution by these methods, a simplified estimation procedure seems to be of practical usefulness. Such a simplified procedure to estimate salt distribution in the soil profile under crop growing field conditions was described by Bresler (1967). The method is essentially a numerical solution of the mass balance equation
(16) when the time increment Δt was set equal to the time
interval between two successive water applications (rain or
irrigation). Furthermore, only the downward flow was considered and was assumed to take place in the range of water
content $(\bar{\theta})$ between saturation and the assumed "field capacity" of the soil. In addition, the amount of water
passing the depth $z \neq 0$ in Δt $[q^j(z)\Delta t^j]$ was estimated by
the difference between the amount of water applied $[Q^j = \bar{q}(0,j)\Delta t^j]$ and the water consumption by the crop from the
soil surface down to the depth of z in Δt^{j-1}. With these
approximations, equation (16) becomes:

$$\int_0^z [c^j(z) - c^{j-1}(z)]\bar{\theta}(z)dz =$$

$$Q^j c_0^{\,j} - [Q^j - \int_0^z E^{j-1}(z)dz]c^{j-\frac{1}{2}}(z) \qquad (17)$$

where Q is the amount of water applied to the soil surface
by irrigation or rain (cm), j is the index of water applied, and E is the amount of water per unit volume of soil
(cm^3 H_2O/cm^3 soil) consumed by the crop during the time interval Δt^{j-1}. Estimation of salt profile from the solution

117

of (17) is possible if $\overline{\theta}(z)$ and $E(z)^{j-1}$ can be evaluated, provided that the initial salt concentration ($c°(z)$), the amount of water applied and their salt concentrations are measured. Evaluation of $E^{j-1}(z)$ may be obtained from the large number of water requirement experiments which have been conducted in many countries (particularly in Israel). Demonstration of the effect of $\overline{\theta}(z)$ on the salt distribution results and on the comparison with measured data are given in Figure 2. The applicability of this estimation method was tested by D. Yaron (1971, and personal communication), who compared many irrigation experiment data with the calculated results, and found to be applicable in many practical cases.

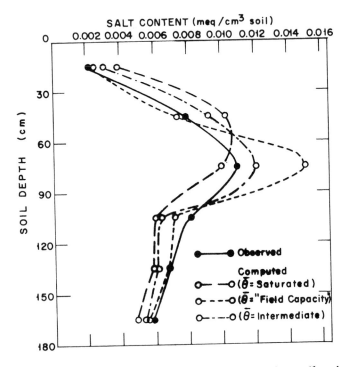

Figure 2. Computed and observed salt distribution data.

OPTIMIZING THE SOIL ENVIRONMENT

Short Run Soil Salinity Control

The problem of soil salinity control, from the point of view of crop yield optimization, may have short-run or long run implications. In the short run the direct effect of salinity on the yield of a specific crop, within a relatively short time period, is considered. The short period involved is usually of the order of one year or one growing season. The short run problem includes reclaiming of initially "saline" soil to optimize the yield of the following crop, and control of soil salinity throughout the crop growing season.

A. Reclamation of Initially Saline Soils by Leaching

Saline soil has been defined as soil "containing enough soluble salts so distributed in the soil profile that they decrease the growth of most plants" (Reeve and Fireman, 1967). Qualitatively, saline soil has been defined as one in which the electrical conductivity of the saturated extract is more than 4 mmho/cm at 25°C and the ESP is less than 15 (Richards, 1954). However, as stated previously, various kinds of crops will react differently to total salt concentration or its composition in the soil solution. Thus, the term saline soil herein shall refer to a soil such that the yield limitation of a given crop of interest grown in it can be attributed to some index of salinity. The reclamation process is aimed at removing this salinity limiting factor so that the crop which follows it will not suffer any yield reduction. Hence, from the point of view of soil reclamation, the definition of saline soil is determined by the characteristics of the crop to be grown.

Reclamation of initially saline soil is essentially a process in which the high-concentration soil solution is displaced by a less concentrated solution. Miscible displacement concepts have been applied to the reclamation of saline soils by leaching soluble salt from them (Biggar and Nielsen, 1967). This concept has been developed and investigated under steady water flow conditions, which are rather unrealistic. However, some consequences of this approach regarding the effect of flow rate and soil water contents on leaching efficiency are applicable. Laboratory miscible displacement results suggested what was found in several field trials that reclaiming saline soil by

leaching is more efficient when the soil is maintained unsaturated and the flow rate is relatively slow (Biggar and
Nielsen, 1962; Nielsen et al., 1965). Similar conclusions
may be drawn from the transient flow computations of Bresler and Hanks (1969).

Although the available theoretical and experimental
analysis of transient water and salt flow processes is as
yet incomplete, it can already be applied in practice for
purposes of reclamation. Here, the estimation procedures
previously described for fallow soils can be used. Knowing
the initial soil conditions and the final desired soil salinity (as determined by the crop to be grown), the quantity-quality relationships of water used in the leaching process can be determined. Leaching efficiency may be improved by regulating the conditions at the soil surface
boundary as for example by artificially decreasing the rate
of water application (Biggar and Nielsen, 1962; Nielsen et.
al., 1965; Bresler and Hanks, 1969). In some cases it is
also possible to increase the leaching efficiency by controlling the bottom boundary conditions by means of artificial drainage. In addition, the reclamation process can be
improved by regulating the $K(\theta)$ function simply by changing
the salt composition and/or concentration of the water applied (Reeve and Bower, 1960; Reeve and Doering, 1966).
Changing the soil hydraulic conductivity by means of soil
conditioning chemicals is also possible. At present, however, these materials appear to be too costly and there is
insufficient information on the persistence of their effect
in the field.

B. Salinity Control During the Crop Growing Period

Control of soil salinity during the growing period of a
given crop depends on knowledge of the crop response pattern to some appropriate soil salinity index. The salinity
index chosen may be the specific concentration of an ion,
or the electrical conductivity of the soil solution (or of
extract obtained from the saturated soil paste). It may
also be defined in terms of the total water suction when
the soil matric suction is also being considered.

Consider first the conception maintained by certain
workers (e.g., Bernstein, 1961a, 1964, 1965; Bierhuizen,
1969) that significant yield response to salinity occurs
only above a certain critical threshold concentration.

Below this threshold value the salinity effect is negligible. This critical salinity index is unique for a given crop and growing conditions. With this conception in mind, equations (15), (16) or (17) can be applied for salinity control purposes.

The systems (15), (16) or (17) can be classified into four major groups: (a) predetermined functions, such as $K(\theta)$, $h(\theta)$, $D(\theta, \overline{V})$, $S(\theta, t, z)$ or $E(z, t)$; (b) predetermined variables (initial conditions) or postreclamation variables $\theta(z, 0)$ and $c(z0)$; (c) man-controlled variables $C_o(t)$, $Q(t) = q(0, t) \Delta t$; and (d) dependent variables, $c(z, t)$, $\theta(z, t)$ and $q(z, t)$. Of particular interest, for salinity control purposes, is the dependence of $c(z, t)$ on the man-controlled variables $Q(t)$ and $C_o(t)$ which uniquely determine the salinity regime of the root zone throughout the growing period (subject to equations (15), (16) or (17), the predetermined functions, and the initial conditions). In other words, control of soil salinity may be obtained by regulating the quantity and/or quality of the water applied.

Let us now calculate those combinations of Q and C_o which yield the same soil salinity index (SS_i) and draw them on a graph (Figure 3), the axes of which shall be Q_t - the cumulative quantity of water applied up to time t, and C_o - the average salt concentration of the applied water. The iso-soil salinity curves demonstrated in Figure 3 are root zone averages calculated by means of equation (17) for a particular time (t) in the growing season. The parametric data were taken from an irrigation experiment performed in the coastal region in Israel. Considering the assumptions underlying equation (17), the shape of each iso-salinity line becomes obvious. The points of minimum C_o represent the cases in which the cumulative amount of water applied, Q_t, is equal to the total amount extracted by the plants in the root zone. Leftward of each point of minimum, only salt accumulation takes place and salinity control due to Q-C_o substitution is irrelevant. Control of soil salinity by the application of excess water for leaching is only possible therefore within the domain lying to the right of these points of minimum.

An application of linear programming and computer simulation modeling to the estimation of iso-soil salinity curves was described by Yaron and Bresler (1970). These methods enable one to consider changes in time and space of the critical threshold values. Thus, variation in the

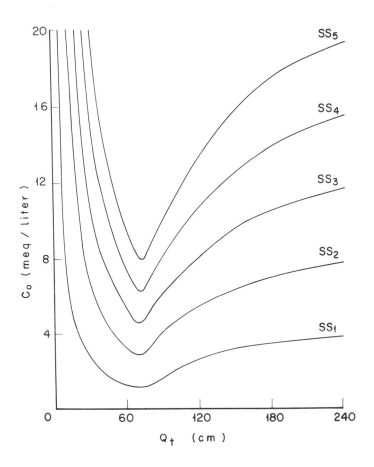

Figure 3. Iso-soil salinity curves (SS_i) as a function of cumulative water application (Q_t) and salt concentration (C_o) of the applied water. The curves were calculated according to Bresler (1967) with data obtained from a citrus irrigation experiment carried out at Shefayim, Israel, on sandy loam soil.

response of the crop during different stages of growth can
be taken into account. The problem may be formulated in
many different ways. To illustrate, let: (a) the amount
of water applied be predetermined at some specific level,
$Q(t)$; (b) the soil salinity index at any time and depth,
$c(z,t)$, be restricted not to exceed prespecified time and
space threshold values, i.e. $c(z,t) \leq C_r(z,t)$. The problem
is to find the maximum level of C_0 which will not violate
the restricted threshold values $C_r(z,t)$ subject to any set
of the finite difference linear form of (15), (16) or (17).
This linear programming formulation consists of a linear
objective function to be maximized subject to a set of linear equalities (the salt distribution estimation methods)
and inequalities (restrictions on the critical threshold
levels). By successively varying the predetermined $Q(t)$
and solving for C_0, a set of Q_t-C_0 combinations is obtained
which maintain the soil salinity index below the prespecified critical values. Another method to solve the same
problem is a computer simulation model. Here the "best"
combination of the man-controlled variables Q_t-C_0 subject
to the same conditions is found by computerized trial and
error (Yaron and Bresler, 1970).

The threshold concentration concept may, in many situations, be of limited use since it represents only a point
on the crop response-salinity function. This point is usually varied when the water regime is changed. An approach
to the empirical estimation of crop response functions to
soil water regime and salinity was presented by Yaron et
al. (1971). They obtained an empirical relationship between crop yield and the total suction (TS) or the actual
soil salinity index which were either measured or estimated.
Available data on crop response function to the combined
effect of soil water regime and salinity, expressed by the
TS index, add an alternative for controlling soil salinity
throughout the irrigation season (Bresler and Yaron, 1971).
Regulating the soil water regime by varying the frequency
of water application and the wetting depth, within the root
zone, will affect the total water suction (TS) by changing
both the actual soil solution concentration and the soil
matric suction. In regions with Mediterranean type climate
(i.e., semiannual alternation of irrigation in dry summers
and rainy winter seasons), it might be more efficient to
control the salinity by regulating the soil water regime
rather than by leaching. Under such conditions, it would

seem better to leach the soil, if necessary, during the rainy season and to control the salinity effect by altering the water regime during the irrigation season. Irrigation frequency (I) is thus an additional man-controlled variable by which the salinity factor can be managed more efficiently (with total soil water suction taken into consideration).

Long Run Soil Salinity Control

In the long run the perennial soil salinity effect must be considered. Usually the long run effects are the time integral of the short run effects. Thus, when the short run effects can be estimated or predicted it is also possible to predict the long run effects simply by integrating all the short period effects over the longer period of interests. An alternative way is to approximate the system by taking average values over relatively large time and space intervals. These are essentially the principles underlying the salt balance and leaching requirement concepts.

A. The Salt Balance Concept

The salt balance concept was first introduced and used by Schofield (1940). In principle the salt balance concept may be defined by equation (16) integrated over the entire root zone (Z) and averaged over the time period (Δt) of interests. Thus if the mass of salt input ($\bar{q}_o \bar{C}_o \Delta t$) exceeds the salt output ($\bar{q}_z \bar{C}_z \Delta t$) from the root zone, then salt is accumulating in the root zone (the LHS integral in (16) is positive) and the trend is undesirable. The usefulness of the salt balance concept as an indicator of salinity trends for the time scale of a year and for a very large area was shown by Wilcox and Resch (1963). From their results it appears that large-scale salt balance data can serve as a guide for determining the necessity of salinity control in irrigation projects.

B. The Leaching Requirement Concept

Leaching requirement has been defined as "the fraction of irrigation water that must percolate through the root zone to keep the salinity of the soil below a specified value" (Richards, 1954). This general definition holds for both short and long run salinity control. However, the working equation for calculating the leaching requirement is more appropriate for long-time than for short-time averages, as it is based on the assumption of steady-state mass flow conditions.

The equation for leaching requirement is a modification of equation (16) for large scale field conditions when steady state salt transport is maintained. This equation, in the notation used in the U.S. Salinity Laboratory Manual (Richards, 1954) is:

$$LR = \frac{D_{dw}}{D_{iw}} = \frac{C_{iw}}{C_{dw}} \qquad (19)$$

where LR is the leaching requirement, D_{iw} and D_{dw} are depths of water entering the soil surface and drained below the root zone, respectively, C_{iw} and C_{dw} are the respective salt concentrations in the irrigation and drainage water (in terms of electrical conductivity or of any specific ion concentration). If D_c is the depth of water representing the total evapotranspiration from the root zone, then $D_{iw} = D_{dw} + D_c$, and therefore the amount of irrigation water to be used when steady conditions are maintained is:

$$D_{iw} = D_c/(1 - LR) = D_c/(1 - C_{iw}/C_{dw}) \qquad (20)$$

In applying equation (19) or (20) to a given salinity control situation one has always to remember the steady state restriction imposed on the system by these equations. Steady state conditions rarely occur in short run irrigated field situations. However, the LR thus defined may be applicable for long-time averages under circumstances of quasisteady state conditions. Under these conditions salinity control is possible by regulating at least two of the variables comprising equations (19) and (20) in order to maintain the steady conditions and the desired concentration throughout the root zone system. Soil drainability or artificial drainage is therefore very important. Drainage is also important in controlling the depth of the water table. This depth should always be kept sufficiently low to minimize upward movement of salt by capillary rise of water from the water table into the root zone and to the soil surface.

Economic Approach to Short Run Salinity Control

Any optimization process of the type considered herein must finally be judged from the point of view of economic efficiency. Optimizing the soil physical environment with respect to soil salinity should be defined by economic analysis of soil salinity control. It is appropriate therefore to conclude this discussion with a brief summary of recent work on the economic analysis of the problem of soil salinity control (Yaron and Bresler, 1970; Yaron et al., 1971; Bresler and Yaron, 1971).

Let us first consider the problem on the basis of the "critical threshold concentration" concept. The economic problem here is to find the least costly combination of water quality (C_o) and quantity (Q) which can ensure that a given critical threshold salinity index (SS_i) is not exceeded. To illustrate the problem let the critical salinity index be represented by the relevant part of the iso-soil salinity line, say SS_5 in Figure 3 which was redrawn in Figure 4. Such a line may be obtained by any of the estimation methods previously described. For the analysis to be realistic, it must be assumed that both Q and C_o are readily controllable, and that the cost of a unit amount of water increases with the decrease of its salt concentration. To derive the optimal Q-C_o combination of the system represented in Figure 4, it is necessary to minimize the total cost of water. To this goal the following Lagrangian expression is formed

$$L = QP_q + (C_s - C_o)P_c Q - \lambda[g(C_o,Q) - SS_5] \qquad (21)$$

where L is the total unit water cost, P_q is the cost per unit amount of water of standard salinity, C_s is the standard water salinity in terms of a salinity index (SI), P_c is the cost of C_o deviating in one water unit by one SI unit, and $g(C_o,Q)$ is the Q-C function in Figure 4. To minimize equation (21) each partial derivative of L with respect to Q, C_o and λ should be equal to zero. In performing this operation we get

$$\frac{dC_o}{dQ} = [P_q + (C_s - C_o)/P_c]/P_c Q$$

$$SS_5 = g(Q,C) \qquad (22)$$

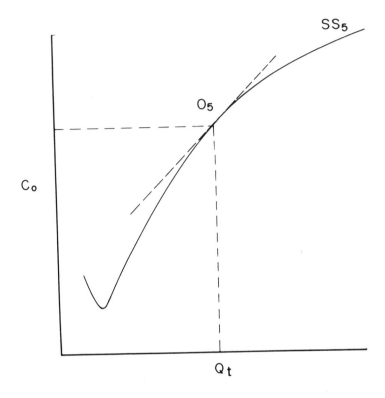

Figure 4. Threshold iso-soil-salinity curve (SS_5 of Figure 3) and determination of the optimal quantity (Q_t)- quality (C_o) combinations.

Equation (22) shows the necessary conditions for the least cost quantity-quality ($Q-C_o$) combination. The least cost combination is illustrated graphically in Figure 4. The curve SS_5 represents the threshold concentration at the $SS = 5$ level and the straight line represents the cost of quantity and quality ratio. The point O_5, in which the straight line is tangent to the iso-threshold salinity line, represents the optimal $Q-C_o$ combination. Note that for any given level of threshold conditions an optimal combination can similarly be derived.

A more general approach to the economic evaluation of salinity control variables has been suggested by Bresler and Yaron (1971). The empirical estimation of functional crop response to salinity and water regime, expressed in terms of total suction, TS, (Yaron et al., 1971) was combined with the estimation of TS (by equation [17]) as a function of the man-controlled variables Q, C_o and irrigation frequency I. Thus, $Y = y(TS)$, and $TS = f(Q, C_o, I)$. Therefore,

$$Y = Y(Q, C_o, I) \qquad (23)$$

where Y is the economic crop yield.

Equation (23) does not include the initial salinity prior to the irrigation season, $c(0,z)$. It was assumed that the value of $c(0,z)$ can be controlled (i.e., by [16] or [17]); otherwise, it must be included in (23). If we assume that Q, C_o and I can be controlled and that regulating I is costless, then the optimal economic combination is derivable from the maximization of

$$R = Y(Q, C_o, I)P_y - [QP_q + (C_s - C)P_c Q] \qquad (24)$$

where P_y is the net price per unit economic yield and the other terms are as defined above. Maximizing R by a similar procedure as before results in the optimal economic management of irrigation for controlling the soil salinity, by regulating the amount, frequency and concentration of the water applied.

References

1. Bernstein, L. and Hayward, H.E. (1958). Physiology of plant tolerance. Ann. Rev. Plant Physiol. 9, 25-46.
2. Bernstein, L. (1961a). Tolerance of plants to salinity. Amer. Soc. Civ. Eng. J. Irrig. Drainage Div. 87(IR4), 1-2.
3. Bernstein, L. (1961b). Osmotic adjustment of plants to saline media. I. Steady state. Amer. J. Bot. 48, 909-918.
4. Bernstein, L. (1964). Salt tolerance to plants. U.S. Dept. Agr. Inform. Bul. 283.
5. Bernstein, L. (1965). Salt tolerance of fruit crops. U.S. Dept. Agr. Inform. Bul. 292.
6. Bierhuizen, J.F. (1969). Water quality and yield depression. Inst. Land and Water Manag. Res., Wageningen,

Tech. Bul. 61.

7. Biggar, J.W. and Nielsen, D.R. (1962). Improve leaching practice - save water, reduce drainage problems. Calif. Agr. 16, 5.

8. Biggar, J.W. and Nielsen, D.R. (1967). Miscible displacement and leaching phenomena. In "Irrigation of Agricultural Lands," pp. 254-274. Agronomy 11, Amer. Soc. Agron., Madison.

9. Bresler, E. (1967). A model for tracing salt distribution in the soil profile and estimating the efficient combination of water quality and quantity under varying field conditions. Soil Sci. 104, 227-233.

10. Bresler, E. (1971). Transient diffusion-convection salt flow in unsaturated soils (unpublished ms.).

11. Bresler, E. (1972). Solute movement in soils, theory and applications. In "An Introduction to Arid Zone Irrigation." Springer Verlag Pub., New York-Berlin (in press).

12. Bresler, E. and Hanks, R.J. (1969). Numerical method for estimating simultaneous flow of water and salt in unsaturated soils. Soil Sci. Soc. Amer. Proc. 33, 827-832.

13. Bresler, E. and Yaron, D. (1971). Soil water regime in economical evaluation of salinity in irrigation. Mimeographed Rept., Volcani Inst., Beit Dagan, Israel.

14. Bresler, E., Heller, J., Diner, N., Ben-Asher, I., Brandt, A., and Goldberg, D. (1971). Infiltration from a trickle source. II. Experimental data and theoretical predictions. Soil Sci. Soc. Amer. Proc. (in press).

15. Freeze, R.A. (1969). The mechanism of natural ground-water recharge and discharge. I. One-dimensional vertical, unsteady, unsaturated flow above a recharging or discharging ground-water flow system. Water Resources Res. 5, 153-171.

16. Gardner, W.R., Mayhugh, M.S., Goertzen, J.O., and Bower, C.A. (1959). Effect of electrolyte concentration and exchangeable sodium percentage on diffusivity of water in soils. Soil Sci. 88, 270-274.

17. Hayward, H.E. (1954). Plant growth under saline conditions. UNESCO Arid Zone Res. 4, 37-71.

18. Hanks, R.J., Klute, A., and Bresler, E. (1969). A numeric method for estimating infiltration, redistribution, drainage, and evaporation of water from soil. Water Resources Res. 5, 1064-1069.

19. Kemper, W.D. and Evans, N.A. (1963). Movement of water as affected by free energy and pressure gradients. III. Restriction of solute by membranes. Soil Sci. Soc. Amer. Proc. 27, 485-490.

20. Kemper, W.D. and van Schaik, J.C. (1966). Diffusion of salts in clay-water systems. Soil Sci. Soc. Amer. Proc. 30, 534-540.

21. Kemper, W.D., Maasland, D.E.L., and Porter, L.K. (1964). Mobility of water adjacent to mineral surfaces. Soil Sci. Soc. Amer. Proc. 28, 16-164.

22. Letey, J. (1968). Movement of water through soil as influenced by osmotic pressure and temperature gradients. Hilgardia 39, 405-418.

23. Molz, F.J. and Remson, I. (1970). Extraction term models of soil moisture use by transpiring plants. Water Resources Res. 6, 1346-1356.

24. Naghskineh-Pour; B., Kunze, G.W., and Carson, C.D. (1970). The effect of electrolyte composition on hydraulic conductivity of certain Texas soils. Soil Sci. 110, 124-127.

25. Nielsen, D.R. and Biggar, J.W. (1961). Miscible displacement in soils: I. Experimental information. Soil Sci. Soc. Amer. Proc. 25, 1-5.

26. Nielsen, D.R. and Biggar, J.W. (1962). Miscible displacement: III. Theoretical considerations. Soil Sci. Soc. Amer. Proc. 26, 216-221.

27. Nielsen, D.R., Biggar, J.W., and Luthin, J.N. (1965). Desalinization of soils under controlled unsaturated flow conditions. Int. Comm. Irrig. Drainage, 6th Congr. Quest 19, 15-24.

28. Ogata, A. (1970). Theory of dispersion in a granular medium. Geological Survey Professional Paper 411-I. U.S. Govt. Printing Office, Washington.

29. Olsen, S.R. and Kemper, W.D. (1968). Movement of nutrients to plant roots. Advan. Agron. 30, 91-151.

30. Passioura, J.B. (1971). Hydrodynamic dispersion in aggregated media. I. Theory. Soil Sci. (in press).

31. Passioura, J.B., Rose, D.A., and Haszler, K. (1970). Lognorm: A program for analysing experiments on hydrodynamic dispersion. CSIRO, Div. of Land Res. Tech. Memorandum 70/6.

32. Perkins, T.K. and Johnston, O.C. (1963). A review of diffusion and dispersion in porous media. Soc. Petroleum Engs. J. 3, 70-84.

33. Porter, L.K., Kemper, W.D., Jackson, R.D., and Stewart, B.A. (1960). Chloride diffusion in soils as influenced by moisture content. Soil Sci. Soc. Amer. Proc. 24, 460-463.

34. Quirk, J.P. (1957). Effect of electrolyte concentration on soil permeability and water entry in irrigation soils. Int. Comm. Irrig. and Drainage, 3rd Congr, R6. Quest. 8, 115-123.

35. Quirk, J.P. and Schofield, R.K. (1955). The effect of electrolyte concentration on soil permeability. J. Soil Sci. 6, 163-178.

36. Reeve, R.C. and Bower, C.A. (1960). Use of high salt water as a flocculant and source of divalent cations for reclaiming sodic soils. Soil Sci. 90, 139-144.

37. Reeve, R.C. and Doering, E.J. (1966). The high-salt water method for reclaiming sodic soils. Soil Sci. Soc. Amer. Proc. 30, 498-504.

38. Reeve, C.R. and Fireman, M. (1967). Salt problems in relation to irrigation. In "Irrigation of Agricultural Lands," pp. 988-1008. Agronomy 11, Amer. Soc. Agron., Madison.

39. Richards, L.A., (ed.)(1954). Diagnosis and improvement of saline and alkali soils. U.S. Dept. Agr. Handbook 60.

40. Schofield, C.S.(1940). Salt balance in irrigated areas. J. Agr. Res. 61, 17-39.

41. Shalhevet, J. (1970). The use of saline water for irrigation. AEAI, ITCC, 2nd Congr., Jerusalem. Reg. No. 19.2, 437-442.

42. Shalhevet, J. and Bernstein, L. (1968). Effects of vertically heterogeneous soil salinity on plant growth and water uptake. Soil Sci. 106, 85-93.

43. Shalhevet, J., Reiniger, P., and Shimshi, D. (1969). Peanut response to uniform and non-uniform soil salinity. Agron. J. 61, 584-587.

44. Van Schaik, J.C. and Kemper, W.D. (1966). Chloride diffusion in clay-water systems. Soil Sci. Soc. Amer. Proc. 30, 22-25.

45. Wadleigh, C.H. and Ayers, A.D. (1945). Growth and biochemical composition of bean plants as conditioned by soil moisture tension and salt concentration. Plant Physiol. 20, 106-132.

46. Wadleigh, C.H., Gandi, H.G., and Kolisch, M. (1951). Mineral composition of orchard grass grown on Pachappa

loam salinized with various salts. *Soil Sci.* 72, 275-282.

47. Wilcox, L.V. and Resch, W.F. (1963). Salt balance and leaching requirement in irrigated lands. U.S. Dept. Agr. Tech. Bul. 1290.

48. Yaron, D. and Bresler, E. (1970). A model for the economic evaluation of water quality in irrigation. *Aust. J. Agr. Econ.* 14, 53-62.

49. Yaron, D., Bielorai, H., Shalhevet, J., and Gavish, Y. (1971). Estimation procedures of response functions of crops to soil water content and salinity. *Water Resources Res.* (accepted for publication).

50. Yaron, D., Weisbrood, M., Stratiner, G., Shimshi, D., and Bresler, E. (1971). Simulation model of water variation in soil. Mimeographed Rept., The Hebrew Univ. Faculty of Agr., Rehovot, Israel.

PROGRAMMING IRRIGATION FOR GREATER EFFICIENCY

M.E. Jensen
Agricultural Research Service
U.S. Department of Agriculture

Introduction

Programming irrigation for greater efficiency implies maximizing water use efficiency. However, though this in itself is an important objective for water deficient areas, the farm manager or owner is more interested in maximizing his net income by optimum use of irrigation water, fertilizers, and other inter-related inputs. Fortunately, if maximum net income is achieved, optimum water use efficiency usually has also been attained. More progress toward greater efficiency can be expected by working toward the goal of greater net income rather than greater water use efficiency, per se.

Evaluations of farm irrigation practices in the western United States during the late 1950's and early 1960's (Tyler et al., 1964; Willardson, 1967) showed little change in irrigation scheduling practices in 25 years (Israelsen, 1944). No single factor related to the system, soil, or crop appeared to be limiting irrigation efficiency. During this same era, irrigation science and technology made significant advancements.

There are two major reasons why irrigation scheduling practices, involving both timing and amount of water applied, have not changed substantially: (1) The needs of managers of irrigated farms and the acceptability of suggested scheduling procedures have not been adequately evaluated; (2) The cost of irrigation water often has not been significant, and indirect costs such as yield reductions caused by delayed irrigations and additional nitrogen requirements created by excessive water applications are not easily recognized or quantified. Also, crop and soil damage costs encountered on lower-lying areas by excessive water use on upper areas are not always borne by the upper-area irrigators.

Irrigation scientists and technologists know how to optimize production by manipulating irrigation practices, but these specialists are not making current irrigation decisions on each farm. Irrigation decisions are made by people who have limited time and training. They do not have the meteorological and crop growth data and forecasts to predict the next date of irrigation so that farm work can be planned accordingly. The modern farm manager can acquire irrigation equipment to apply the amount of water needed when it is needed, but he still must decide "when" and "how much" water to apply. Predictions of needed irrigation several days in advance are essential for planning other farm work and for completing the required irrigations when the capacity of the irrigation system or allotted flow is limited.

Basically, current estimates and predictions of irrigation needs are not available to most managers of irrigated farms, or at least they are not available in a form that can be used. Irrigation scheduling is a decision-making process requiring current information, trends, projections, and alternatives much the same as required by managers of large industries. The modern farm manager needs and wants a continuing service that gives the present soil water status on each of his fields, predicts irrigation dates, and specifies the amounts of water to apply on each field. He could also use predictions of adverse effects, such as the effects of delaying an irrigation for several days, or perhaps terminating irrigations, on the yield of marketable products. This information would increase the manager's skills in making better and more profitable irrigation decisions.

An irrigation management service is often needed to supplement practical irrigation experience with irrigation science on a day-to-day basis. The potential economic returns to the farm manager or owner can exceed the cost of this service severalfold. Moreover, the increase in net income to the farmer and greater water use efficiencies are not the only benefits to be derived from this service. The recipients of the irrigation forecasts have found the data to be useful in planning other farm work, such as cultivation and spraying for insect control. In semihumid areas, it might enable farmers to begin irrigating their crops in sufficient time to cover all of the fields before soil

moisture deficits become severe. With the current high interest and enthusiasm by the managers of irrigated farms in
the United States for a service such as this, and the establishment of new irrigation service companies or the addition of irrigation scheduling services to existing companies, it is expected that a significant portion of the irrigated farms in the United States will be scheduled using
procedures such as described in this chapter within the
next few years.

Alternative Methods of Improving Irrigation Programming

There are currently four general approaches to improving irrigation scheduling practices.

A. Irrigation Programming Instruments

Tools and instruments, such as tensiometers, soil water
blocks, evaporation pans, and soil sampling augers and
tubes have long been recommended or supplied to farm managers with instructions for their use. These instruments
are often supplemented with guides to interpreting the results obtained. This has been the general approach taken
by most irrigation technologists during the last three decades to improve irrigation water management, and there are
numerous technical publications and manuals on this subject.
New and better soil water instruments are technical aids to
the solution of the problem, but do not in themselves ensure a practical solution, as evidenced by continued poor
irrigation practices in many areas.

B. Irrigation Management Services Using Instruments

An irrigation management service by trained service
groups or private firms using some of the tools mentioned
in the previous section is a practical alternative and is
being used to a limited extent in some areas of the western
United States. Average evapotranspiration data determined
experimentally in the area may be used for predicting the
date of the next irrigation. Irrigation management services based on soil sampling are in use in Arizona and Washington, U.S.A. (Franzoy and Tankersley, 1970; Marshall,
1971). In general, a service involving soil sampling has
not been widely adopted because farm managers are not

always aware of the need and benefit of good irrigation water management. Services using tensiometers are more common, especially with high-value crops, but these instruments must be read frequently, which involves significant travel and labor costs.

C. General Irrigation Forecasts

Generalized forecasts of irrigation needs for common crops and the major soils in the area based on local or regional experimental data are frequently used in areas needing only infrequent supplemental irrigations. This approach is economical, but generally requires the farm manager to interpret the forecasts and monitor his own fields to verify the predictions (Jensen, 1969). This type of service has also been adapted to arid conditions. For example, a general forecast service is being developed in southern Idaho with the predictions provided by the U.S. Bureau of Reclamation and the A & B Irrigation District of the Minidoka Project (Brown and Buchheim, 1971).

D. Field Scheduling and Monitoring

Irrigations are predicted for each field utilizing meteorological data combined with soils, crop, and experimental data, but these predictions are supplemented with a field monitoring service. This alternative has been tried and is readily accepted by the managers of irrigated farms (Jensen et al., 1970; Franzoy and Tankersley, 1970).

Irrigation Management Service Requirements

The scientific knowledge to estimate and predict the depletion of soil water by evapotranspiration has been the subject of world-wide research since the late 1940's, but estimating or predicting soil water depletion solves only part of the problem. Some of the additional requirements that are essential in providing an irrigation management service are described in this section.

A. Technical Competence

A service group such as a governmental agency, irrigation district, or a private firm must have the necessary technical competence to collect and interpret essential basic data and develop the predictions. There are advantages in private groups providing this service, as the private consultant can supply other personalized management services to the farm manager. Many private firms, for example, have the skills in other areas such as fertilizer management, pesticide control, cost accounting, and general agronomy. Private service groups should develop self-regulating standards of staff competence to assure dependable recommendations for irrigation clients. If this is not done, state licensing as required for other professional services, will soon be required.

B. Economic Feasibility

Unwarranted accuracy or complexity should be avoided in order for a service to be self-supporting. The costs of an irrigation scheduling service should be economical relative to other irrigation operation and maintenance costs. This means that expensive instrumentation and detailed measurements on each farm probably will not be feasible. Also, it means that some basic meteorological data cannot be used because complicated instrumentation, data processing, and instrument servicing would be required. The service company or agency must utilize meteorological data that are easily obtainable from existing weather stations, or that can be obtained with reliable instruments that function throughout the season with few or minor mechanical problems.

C. Basic Farm Data and Records

The servicing group must collect background information on each field, including soils data, the crops to be grown, past cropping history where soil fertility is involved, characteristics of each existing irrigation system, and current practices of each farm manager. It must also maintain essential records for each farm and field involved. If a generalized service is provided, the amount of record keeping involved is substantially reduced, but the accuracy of generalized forecasts for each field may be lower, especially if field monitoring by skilled technicians is not provided.

D. Communications

When operating on a field-by-field basis, a two-way communication system must be an important part of the program and has been one of the major problems encountered by service groups. Complete reliance on a mail service has not been fully satisfactory because of the time lag occasionally encountered. Also, farm managers frequently neglect to promptly provide the desired feedback information, such as the date a field has been irrigated, which would be valuable even to a company that relies on soil sampling. A telephone service for collecting the feedback information has been tried. However, this procedure has not been satisfactory because the farm manager frequently is not in when called, or the service company office may be closed when the farm manager chooses to make his own call. Private service groups in Southern Idaho are considering installing automatic telephone recording systems so that the farm manager can call the service center and provide feedback information at his convenience.

The Salt River Project at Phoenix, Arizona, is contemplating using remote computer terminals connected by telephone network to a central computer. The remote terminals will be located in outlying field offices where the farm managers regularly stop to order irrigation water. The terminal operator can reach the central computer files and obtain an updated printout for the manager's farm without delay. The printout lists the predicted irrigation dates for each of the fields involved in the scheduling program. This arrangement is desirable when an irrigation district is providing the service since it also controls water distribution and maintains records of irrigation water deliveries. As an alternative, the Salt River Project is evaluating the use of a time-sharing terminal in the main office. with data available to the field by telephone.

E. Verification and Field Inspection

The success of an irrigation management service will, to a large extent, depend on periodic monitoring or soil samplings of the fields to verify the adequacy of previous irrigations and the accuracy of predicted irrigation dates. This part of the service can be provided by a technician experienced in irrigation water management, or the soil

sampler in a soil sampling program. Preferably, one technician serves a specific group of farm managers so that the
manager develops a personal rapport with his service technician. He looks to the technician or his supervisor as an
expert in irrigation water management, and one who can also
provide guidance on other aspects of surface or sprinkler
irrigation. The Salt River Project at Phoenix, Arizona
utilizes technicians in this manner with one technician
serving about 20 farms. He visits each farm weekly, and if
he encounters questions that he cannot answer, his vehicle
is equipped with a two-way radio so that he can call his
supervisor or a specialist in the central office to obtain
a specific recommendation.

Meteorological Data Limitations

Energy balance and, in some cases, mass transfer measurements have become reliable and accurate methods of determining daily evapotranspiration in experimental studies.
However, the cost, technical skills, and data processing
required effectively prohibit these techniques on a continuing daily basis even for a single reference field within a
project. Instead, estimates of "potential evapotranspiration," or evapotranspiration that occurs with a well-watered reference crop like alfalfa with 30 to 50 cm of
growth within the area, are adequate. Also, if a combination equation is used, only daily values of meteorological
data are needed to provide the necessary accuracy of $\pm$ 10
to 15% on a daily basis. The accuracy for 5- to 20-day
periods will be better if the errors are random.

There are also other problems in obtaining the necessary meteorological data. In some areas, a service company
must establish its own weather station. In the Western
United States, a weather station centrally located within
an irrigated project of 50,000 to 100,000 hectares can provide adequate data to compute the daily evaporative demand.
Observer skill and instrument calibration are major problems encountered in the collection of meteorological data.
Table 1 indicates the degree of skills that should be considered when collecting various meteorological data. This
table is based on personal experiences of encountering er-
roneous data where meteorological stations are not properly
maintained, or where meteorological readings are not taken
by trained technicians. Some of the refined measurements

required for energy balance computations may actually require a scientist's daily attention. Since this degree of
skill is not practical for most irrigation scheduling services, procedures that require daily, precise meteorological measurements are restricted.

TABLE 1
Technical Skills Required to Obtain Reliable Daily
Meteorological Data on a Routine Basis without Daily
Supervision by a Research Engineer or Scientist.
(The x's denote an acceptable level of training.)

Meteorological parameters	Training required		
	Observers	Technicians	Meteorologists
Air temperatures			
Max-min thermometers	x	x	x
Thermographs	x*	x	x
Humidity (dew point)			
Sling psychrometer	x*	x	x
Hygrograph	-	x*	x
Wind (daily run)			
Integrating anemometer	x*	x	x
Solar radiation (daily)			
Actinograph	-	x†	x*
Integrating unit (sensor-integrator)	-	x†	x*
Sunshine hours	-	x*	x
Net radiation			
Integrating unit	-	-	x*
Evaporation			
Standard pans	-	x*	x

* Periodic calibration, maintenance and instruction needed.
† Periodic calibration, maintenance and instruction by a
  trained meteorologist, research engineer or scientist.

OPTIMIZING THE SOIL ENVIRONMENT

Effective Use of Regional Experimental Data

In the U.S.A., the amount of research data available
for most large irrigated areas is often extensive and represents a large investment of private and public funds.
The interpretation of experimental data for optimum irrigation water management requires skills in the science of
agronomy and plant-soil-water relations. Similarly, the
collection and interpretation of soils data available from
previous experiments requires practical knowledge of soil-
water principles. Workshops are frequently conducted for
farm managers. Similar but more technical workshops are
needed for the professional staff members of service companies. There are some circumstances in which knowledge of
soil-water storage is of no avail. These usually occur
when the irrigation system can apply only a limited amount
of water per irrigation, less than the generally allowable
soil water depletion for a given soil and crop, and when
expected rainfall is insignificant. In such cases, the actual allowable depletion of soil water between irrigations
will be largely determined by the irrigation system and not
by soil characteristics.

USDA-ARS-SWC Irrigation-Scheduling Computer Program

A. History of Development

The USDA computer program was developed cooperatively
with farm managers and service groups as a tool for providing managers of irrigated farms with scientific estimates
of irrigation needs for each field. This approach is not
the only solution to the problem, but it is one that has
gained rapid acceptance. The computer program requires
limited input data and uses simple, basic equations so that
each can be replaced as more accurate relationships are developed. The principles and procedures involved are described in several recent publications (Heermann and Jensen,
1970; Jensen, 1969; Jensen and Heerman, 1970; Jensen et al.,
1971). A summary of the computer program is given in the
next section.

The basic components of the irrigation management service were evaluated in 1966 and 1967 in southern Idaho.
The computer program was evaluated in 1968 and 1969 on
about 50 fields in Idaho and a similar number in Arizona by

the Salt River project (Franzoy and Tankersley, 1970).
During 1970, a number of service groups and companies
gained experience in the use of this general concept of irrigation scheduling. The Salt River Project at Phoenix,
Arizona, for example, has used the program for three years.
After the first two years, the original program was revised,
retaining the basic components, but the input-output data
and format and some of the crop curves were changed to fit
local facilities and crops (Franzoy and Tankersley, 1970).
The irrigation predictions should be updated twice a week
when shallow-rooted crops are involved or when evapotranspiration rates are high. Weekly updating suffices for most
field crops.

The U.S. Bureau of Reclamation has modified the program
to provide general irrigation forecasts for major crops in
an area. This service was made available to southern Idaho
for 54 farm operators in 1970. A field-by-field scheduling
and monitoring service was provided for 68 fields on 15
farms within the same area (Brown and Buchheim, 1971). The
general forecasts were updated once a week and distributed
to cooperators who provided their own field monitoring. An
agricultural technician made weekly visits throughout the
irrigation season on the 15 farms on the field-by-field
program. Additional changes have been made for 1971 which
include both the General Irrigation Forecasts and Field
Scheduling and Monitoring Methods. These methods will be
used in 1971 on two irrigation districts in southern Idaho
with field monitoring provided by trained technicians on
each farm.

An agriculture technology company in McCook, Nebraska
is using a modified version of an earlier program in Nebraska and Kansas (Corey, 1970). This company is also
planning to add fertilizer management to its service in
1971.

Approximately 8,000 hectares were scheduled in 1970 by
various groups in Idaho, Nebraska, and Arizona. An estimated 50,000 hectares will be scheduled in 1971, with nearly half of the area involving cotton on the Salt River Project. Other private companies are adding this service to
their regular farm management services in Idaho, Washington,
Kansas, and California. The Bureau of Reclamation is also
expanding its irrigation management program to other areas
such as the Central Valley of California, and the Rio
Grande Valley of Texas.

The concept of scheduling irrigations using climatic data is not new. Das (1936) suggested using climatic data to control irrigations in the 1930's. The concept received more attention following the publications of Penman (1948, 1952) and Thornthwaite (1948). In 1954, Baver stated:

> The meteorological approach to irrigation has the advantage of simplicity of operation when compared with methods based upon measurement of soil moisture changes. If it is proved satisfactory, the costs of using this system would be relatively small. Undoubtedly, new techniques will be developed that will give an integrated measure of daily temperature, sunshine and solar energy. When such methods are available, meteorological data can be correlated better with evapotranspiration.

Many others have since discussed this approach (Baier, 1957, 1969; Pierce, 1958, 1960; Pruitt and Jensen, 1955; Rickard, 1957; van Bavel, 1960; van Bavel and Wilson, 1952). However, prior to 1965 this method had not been adapted for general practical use or tested extensively in the United States. The Penman equation is often referred to as being too complex for practical use (Fitzpatrick and Cossens, 1965). However, with modern low cost computers, there is no justification in using less refined methods of computation when the meteorological data are easily obtainable.

B. Operating Costs

The current costs of this service range from $4.00 to $10.00 (U.S.) per hectare. The lower cost is for large acreages, crops requiring less frequent monitoring, and short-season crops.

C. Operational Steps

The program first estimates daily evaporative flux from a well-watered reference crop like alfalfa with 30 to 50 cm of top growth, E*. The basic meteorological data required for this estimate are: (1) daily maximum and minimum air temperatures; (2) daily solar radiation; (3) average daily dewpoint temperature or dewpoint temperature observed at or near 8:00 A.M.; and (4) daily wind run at a known height, preferably in an open area over a surface that does not change significantly in roughness or displacement height during the growing season. Prior to 1971 (Jensen et al., 1971), the combination equation as presented by Penman (1963) was used in the following format:

$$E^* = \frac{\Delta}{\Delta + \gamma} (R_n + G) + \frac{\gamma}{\Delta + \gamma} (15.36)(1.0 + 0.0062u_2)(e_s - e_d) \tag{1}$$

where Δ is the slope of the saturation vapor pressure-temperature curve, (de/dT), mb $°C^{-1}$; γ is the psychrometric constant (0.66 mb $°C^{-1}$ at 20°C and 1 bar pressure); e_s is the mean saturation vapor pressure in mb (mean at maximum and minimum daily air temperatures); and e_d is the saturation vapor pressure at mean dewpoint temperature in mb; u_2 is total daily wind run in km day^{-1} at a height of 2 m; R_n is net radiation in cal cm^{-2} day^{-1}, and G is heat flux from the soil in cal cm^{-2} day^{-1} (negative when the heat flux is to the soil). The parameters $\Delta/(\Delta + \gamma)$ and $\gamma/(\Delta + \gamma)$ are mean air temperature weighting factors whose sum is 1.0.

Since the percent of sunshine or degree of cloud cover, normally used to estimate net longwave radiation, are generally not available, or are qualitative rather than quantitative, procedures were developed for estimating net radiation using observed daily solar radiation, R_s, relative to solar radiation that would normally be expected on that day if there were no clouds, R_{so}. Cloudless-day values can be obtained from various tables such as those by Fritz (1949) or Budyko (1956) or by plotting observed solar radiation to obtain an envelope curve through the high points. Net radiation in cal cm^{-2} day^{-1} is then estimated as follows:

$$R_n = (1 - \alpha) R_s - R_b \tag{2}$$

where α is the mean daily shortwave reflectance or albedo, and R_b is the net outgoing longwave radiation. An albedo of 0.23 is currently used in the program. R_b is estimated as follows:

$$R_b = (a_1 \frac{R_s}{R_{so}} + b_1) R_{bo} \tag{3}$$

where R_{bo} is the net outgoing longwave radiation on cloudless days. The constants a_1 and b_1 derived with data from

Davis, California[1] are 1.35 and -0.35, respectively, and for Idaho 1.22 and -0.18, respectively (Wright and Jensen, 1971).

The net outgoing longwave radiation on cloudless days is estimated using:

$$R_{bo} = (a_2 + b_2\sqrt{e_d})(11.71 \times 10^{-8}) \frac{T_2^4 + T_1^4}{2} \qquad (4)$$

where e_d is the saturation vapor pressure at mean dewpoint temperature in mb; 11.71×10^{-8} is the Stefan-Boltzmann constant in cal cm^{-2} day^{-1} °K^{-4}; and T_2 and T_1 are the maximum and minimum daily air temperatures, respectively, in °K. The constant, a_2, formerly used was 0.32. However, recently a slight variation in the constant a_2 improved the estimates of net radiation (Wright and Jensen, 1971):

$$a_2 = 0.325 + 0.045 \sin [30(M + D/30 - 1.5)] \qquad (5)$$

where M is the month, 1-12, and D is the day, 1-31. The constant, b_2, now used is 0.044, which is the average of the value obtained by Goss and Brooks (1956) in California, 0.040, and the value obtained by Fitzpatrick and Stearn (1965) in Australia, 0.049. These constants in the Brunt equation for effective atmospheric emittance were found to be more suitable for arid conditions than those originally proposed by Penman.

Daily soil heat flux, G, is estimated with a simple empirical equation using the difference between mean daily air temperature and the average temperature for the three previous days. This component is currently being revised. For practical purposes, it can be assumed to be zero under a crop with a full cover, such as alfalfa.

Estimates of daily evaporative flux for the reference crop, E*, in cal cm^{-2} day^{-1}, are converted to depth equivalent, E_{tp}, using 585 cal g^{-1} as the heat of vaporization. Evapotranspiration for a given agricultural crop, E_t, is estimated from the daily reference evaporative flux as follows:

$$E_t = K_c E_{tp} \qquad (6)$$

where K_c is a dimensionless coefficient similar to that

---

[1] W.O. Pruitt, personal communication.

proposed by van Wijk and De Vries (1954) and Makkink and van Hermst (1956). This crop coefficient represents the combined relative effects of the resistance of water movement from the soil to the various evaporating surfaces and the resistance to the diffusion of water vapor from the surfaces to the atmosphere, as well as the relative amount of radiant energy available as compared to the reference crop (Jensen, 1968).

$$K_c = \frac{R_n + G + A}{R_{no} + G_o + A_o} \tag{7}$$

where A is the sensible heat flux to (-) or from the air (+) and G is the sensible heat flux to (-) or from the soil (+); and R_n is net radiation to the crop-soil surface (+). The subscript o designates concurrent values for the reference crop in the immediate vicinity (in this case, alfalfa). The major term affecting K_c is A, which is usually negative on a daily basis for crops with small amounts of leaf area in arid zones, but it may be positive for alfalfa. The crop coefficient can also be expressed in terms of mean daily Bowen ratios and net radiation (Jensen, 1968).

Typical examples of the effects of growth stage on the crop coefficient where soil water is not limiting have been presented for grain sorghum by Jensen (1968), and for corn by Denmead and Shaw (1959). More recently, there have been numerous publications presenting observed relative rates of evapotranspiration as compared to an estimate or measurement of the evaporative potential (e.g., Ritchie, 1971; Ritchie and Burnett, 1971). Mathematical models of K_c based on leaf area index, leaf stomatal resistance to diffusion of water vapor, soil water, and other relevant parameters, may be used in place of experimental values when they become practical for use with limited data.

Since the actual crop coefficient K_c is generally influenced by the wetness of the soil surface, it is automatically adjusted in the computer program as follows:

$$K_c = K_{co} K_a + K_s \tag{8}$$

where K_{co} is the expected crop coefficient based on experimental data where soil water is not limiting and normal plant densities are used; and K_a is the relative coefficient related to available soil water. Currently K_a is assumed to be proportional to the logarithm of the percentage of remaining available soil moisture (AM):

$$K_a = \ln (AM + 1)/\ln 101$$

K_s is the increase in the coefficient as a result of the soil surface being wetted by irrigation or rainfall. The maximum value of K_c normally will not exceed 1.0 for most crops. Currently, the values for K_s for the first, second and third day after a rain or irrigation have been taken as follows: $(0.9 - K_{co})$ 0.8; $(0.9 - K_{co})$ 0.5; and $(0.9 - K_{co})$ 0.3, respectively.

Soil water depletion after an irrigation is calculated as follows:

$$D = \sum_{i=1}^{n} (E_t - R_e - I + W_d)_i \qquad (9)$$

where D is the depletion of soil water (after a thorough irrigation D = 0); R_e is effective rainfall (excluding run-off); I is irrigation water applied, W_d is drainage from the root zone or upward movement from a saturated zone; and i = 1 for the first day after a thorough irrigation when D = 0. The terms on the right hand side of the equation are daily totals in centimeters.

The date of the next irrigation is predicted by the remaining soil water that can be safely depleted and the current average E_t.

$$N = \frac{D_o - D}{\overline{E}_t} \qquad (10)$$

$$N = 0 \text{ for } D > D_o$$

where N is the estimated days to the next irrigation if no rain occurs; D_o is the current optimum depletion of soil water in cm; D is the estimated depletion to date in cm; and $\overline{E}_t$ is the current mean rate of E_t in cm day^{-1}. The magnitude of D_o will vary with each crop and field, and will be partially dependent on the irrigation practice used. The amount of water added to a given soil by furrow irrigation during a 12-hour irrigation, for example, will be affected by furrow slope, stream size, etc.

The amount of water required for the next irrigation, W_I, is calculated by dividing D_o by the attainable efficiency for the irrigation system.

147

Adjustments for leaching can be made, if necessary. At this time periodic monitoring of the salt concentration in the soil is recommended. If the amount of water applied and its salt concentration are known, then automatic adjustments for leaching can be added to the program.

D. Input Data

There are three categories of input data required which should be provided by the service groups working with the farm managers.

Basic Data: Basic data consist of the regional constants for the E* equations, taking into account the differences in height of wind measurements, and the crop-soil system data for each field. The latter item involves the farm name, crop code number, crop-field identification, planting date, estimated effective cover date, estimated harvest date, estimated overall irrigation efficiency for each field based on the system being used, and the maximum amount of soil water that could be depleted by evapotranspiration for each crop. The maximum depletion by evapotranspiration is estimated as the difference between a soil water content about four days after irrigation of a soil that is about 60 to 90 cm deep and has been covered to prevent evaporation (Miller, 1967), and the soil water content reached when the given crop with a fully developed root system is allowed to grow without irrigation until growth ceases. If the amount of water applied is known, then a function for W_d can also be used in equation (9).

Current Meteorological Data: Current meteorological data required for each region are: daily minimum and maximum air temperatures, daily solar radiation, daily dewpoint temperature, and total daily wind run for each day since the last computation date. An optional brief weather forecast can be included for each region, and a coefficient adjusting the expected E* for the next five days either upward or downward can be included based on current forecasts.

Current Field Data: Current data for each field are: date of the last irrigation, the allowable soil water depletion at the present stage of growth (this can be included in the program), and the rainfall and/or irrigation amount and its date of occurrence if it falls within the present computation period.

OPTIMIZING THE SOIL ENVIRONMENT

E. Output Data

The output data can be modified by the service groups
to suit their operating procedures and facilities. A typical example of the output received by a farm manager is illustrated in Computer Output Sheet A.

A typical example of a portion of the output provided
the farm managers receiving generalized forecasts of irrigation needs is given in Computer Output Sheet B.

F. Recent Modifications

Calibration of E* Equations: Under arid conditions,
the constants in the Penman equation tend to underestimate
E* during high advective conditions (Jensen et al., 1971;
Rosenberg, 1969). Under these conditions, the magnitude of
the aerodynamic portion of the combination equation is significantly larger than in semihumid areas. This can best
be illustrated by use of several examples and a derivation
similar to that presented by van Bavel (1966) as follows:

$$E^* = \frac{\Delta}{\Delta + \gamma} (R_n + G) + \frac{\gamma}{\Delta + \gamma} L B_v d_a \qquad (11)$$

where L is the latent heat of vaporization; B_v is a transfer coefficient, g cm^{-2} min^{-1}; and d_a is the saturation vapor pressure deficit of the air, mb. By algebraic manipulation of equation (11) and the energy balance equation
given in the next paragraph, it can be shown that:

$$L B_v d_a = E^* + \frac{\Delta}{\gamma} A \qquad (12)$$

For the examples, assume the following energy balance
equation and meteorological conditions:

$$E^* = (R_n + G) + A$$
$$T = 26°C$$
$$P = 1000 \text{ mb}$$

Under these conditions,

$$\Delta/\gamma = 3, \frac{\Delta}{\Delta + \gamma} = 0.75, \text{ and } \frac{\gamma}{\Delta + \gamma} = 0.25$$

149

Computer Output Sheet A

REGION:BURLEY-TWIN FALLS

FARM:JOHN DOE EXAMPLE   DATE OF COMPUTATION:JULY 28, 1969

CROP-FLD	COEF	SOIL MOISTURE DEPLETION TO DATE	OPTIMUM	RATE	IRRIGATIONS:LAST,NEXT & AMOUNT LAST	: RAIN=0	: W RAIN	: AMOUNT CM
W WHEAT	0.10	3.5	15	0.1	JUL 10	NONE 0	NONE 0	0
BEANS	1.01	1.8	5	.7	JUL 26	AUG 3	AUG 3	9
PEAS	0.10	11.0	11	.1	JUL 07	AUG 3	AUG 3	17
POTATOES	0.90	3.0	5	.6	JUL 24	AUG 1	AUG 1	7
SUG BEETS	0.90	3.7	8	.6	JUL 23	AUG 4	AUG 4	12
CORN	0.93	2.5	8	.6	JUL 25	AUG 5	AUG 5	12
ALFALFA	0.67*	12.3	18	.4	JUL 11	AUG 9	AUG 9	27
PASTURE	0.87	0.0	8	.6	JUL 28	AUG 10	AUG 10	12

PROBABLY RAIN NEXT TWO WEEKS = 0.1 CM

FORECAST:PARTLY CLOUDY & WARM

*The low coefficient for alfalfa reflects the effects of a recent cutting.

150

Computer Output Sheet B

CONSUMPTIVE USE AND SCHEDULING INFORMATION*

JULY 14, 1970 – A&B IRRIGATION DIST -2- SILT LOAM SOIL

DATE OF NEXT IRRIG AT LEFT IF DATE OF LAST IRRIG IS SHOWN BELOW

NEXT IRRIG	SUGBTS	POTATOES	S. GRAIN	BEANS	PEAS	CORN
JUL 14	JUN 28	JUL 5	JUN 26	JUL 5	JUL 3	JUN 30
JUL 15	JUN 29	JUL 7	JUN 27	JUL 6	JUL 4	JUL 1
JUL 16	JUL 1	JUL 8	JUN 29	JUL 7	JUL 5	JUL 2
JUL 17	JUL 2	JUL 10	JUN 30	JUL 9	JUL 6	JUL 4
JUL 18	JUL 2	JUL 11	JUN 30	JUL 10	JUL 7	JUL 5
JUL 19	JUL 3	JUL 12	JUL 1	JUL 11	JUL 8	JUL 6
JUL 20	JUL 4	JUL 12	JUL 2	JUL 12	JUL 9	JUL 7

*Computer Output Sheet 1970 – "General" Method. (Adapted from Figure 3, Brown and Buchheim [1971]).

Example (1) advective conditions:

$$(R_n + G) = 370 \text{ cal cm}^{-2} \text{ day}^{-1}$$
$$A = 130 \text{ cal cm}^{-2} \text{ day}^{-1}$$
$$E^* = 500 \text{ cal cm}^{-2} \text{ day}^{-1}$$

From equations (11) and (12), the following results are obtained:

$$E^* = 0.75(370) + 0.25[3(130) + 500] = 277.5 + 0.25(890) =$$

$$500 \text{ cal cm}^{-2} \text{ day}^{-1}$$

Example (2) sensible heat transfer to the air:

$$(R_n + G) = 370 \text{ cal cm}^{-2} \text{ day}^{-1}$$
$$A = -70 \text{ cal cm}^{-2} \text{ day}^{-1}$$
$$E^* = 300 \text{ cal cm}^{-2} \text{ day}^{-1}$$

When these data are substituted into equations (11) and (12), the following results are obtained:

$$E^* = 0.75(370) + 0.25[3(-70) + 300] = 277.5 + 0.25(90) =$$

$$300 \text{ cal cm}^{-2} \text{ day}^{-1}$$

These two examples clearly illustrate that the aerodynamic term becomes significantly larger when there is warm air advection. Consequently, the aerodynamic term must be more accurate under arid conditions where warm air advection is common (a negative mean daily Bowen ratio) than in semihumid areas where sensible heat is generally transferred to the air from the crop surface (a positive mean daily Bowen ratio).

Since the Penman equation underestimates E*, Wright and Jensen (1971) developed the following empirical coefficients for a linear wind function in equation (1) using lysimeter data and local U.S. Weather Service meteorological observations:

$$f(u) = 0.75 + 0.0114u_2 \qquad (13)$$

where u_2 is the windspeed measured at a height of 2 m in km day^{-1}. This wind function gives essentially the same results at moderate windspeeds as does the van Bavel aerodynamic term $(7.15u_z/[\ln(z/z_o)]^2$ in place of $15.36(1.0 + 0.0062u_2)$ in equation (1), providing that the roughness parameter, z_o, is no greater than 1 cm.

Probability of Rainfall: Heermann and Jensen (1970) modified the computer program to include expected rainfall in predicting the date of the next irrigation. They also modified the estimates of expected potential evapotranspiration when the irrigation date is more than five days in the future, as follows:

$$E[E_{tp}] = E'_{tp} \exp \left[-\left(\frac{t - t'}{\Delta t}\right)^2\right] \qquad (14)$$

where $E[E_{tp}]$ is the expected value of potential evapotranspiration at a given day t (in Julian days); t' is the Julian calendar day when the maximum mean potential evapotranspiration, E'_{tp}, occurs (about July 15 to July 25 in the northern hemisphere); and Δt is the days before and the days after t' when $E[E_{tp}] = 0.37E'_{tp}$. Normally a different value of Δt is used for $t \leq t'$ as compared to $t > t'$. The major advantage of this procedure is that expected potential evapotranspiration is represented by a simple 3-parameter equation using E'_{tp}, t', and Δt. For example, in southern Idaho $E'_{tp} = 0.81$ cm, $t' = 206$, and $\Delta t = 150$ for $t \leq t'$ and 93 for $t > t'$.

Two expected dates of the next irrigation are then calculated: one assuming no rain, and one assuming 50% probable rainfall. Daily depletion is first calculated until $D \geq D_o$ using equations (9) and (15).

$$(E_t)_i = (K_c E[E_{tp}])_i \qquad (15)$$

Then this date is extended by the expected rainfall during the intervening period. If the next irrigation is predicted within the next five days, then $E[E_{tp}]$ is increased or decreased by a current forecast coefficient.

Effective Field Capacity: A better estimate of "effective field capacity" can be obtained by estimating cumulative drainage expected between irrigations. Estimates of the cumulative drainage, W_D, between irrigations can be obtained using the following equations that are based on the empirical relationship of water content and time for a soil that is draining:

$$W = cW_o t^{-m} \qquad (16)$$

153

where W is the water content in cm; W_o the water content
when $t = 1$; m is a constant for a soil; and c is a dimen-
sional constant, t^m (Ogata and Richards, 1957). Accordingly,

$$\frac{dW}{dt} = -mW \left(\frac{W}{cW_o} \right)^{\frac{1}{m}}$$

(17)

The cumulative drainage can be calculated in a manner similar to that proposed by Wilcox (1962):

$$W_D = \sum_{i=1}^{\infty} m[W_{i-1} - (E_t)_i] \left| \frac{W_{i-1} - (E_t)_i}{W_o} \right|^{\frac{1}{m}}$$

(18)

where W_D is drainage, cm; i is the number of the day after
irrigation; and E_t is the evapotranspiration for the day in
cm. The most representative time to sample a soil that has
been covered to prevent evaporation after a thorough irrigation can be obtained by integrating equation (17) between
the limits of W_o and W_D.

$$t = \left| \frac{cW_o}{W_o - W_D} \right|^{\frac{1}{m}}$$

(19)

Since dW/dt rapidly $\rightarrow 0$ using equation (17), about 5 to 20
days of calculations are needed before $dW/dt < 0.01$ cm
day^{-1}. Example calculations based on data from southern
Idaho and unpublished data from D.E. Miller (USDA-ARS-SWC,
Prosser, Washington) are summarized in Table 2. In these
examples, E_t was assumed to be 0.8 cm day^{-1}. In detailed
laboratory measurements, Miller and Aarstad (1971) found
that sampling 3.5 days after irrigation slightly underesti-
mated the available water for the 0- to 70-cm depth, and
sampling at 5.5 days closely represented the available water in the 0- to 120-cm depth with $E_t = 0.81$ cm day^{-1}.
Equation (16) can be used to estimate drainage because the
hydraulic gradient near the lower part of the profile at
higher soil water levels is similar with or without E_t oc-
curing.

154

TABLE 2

Estimated Time to Sample a Soil to Determine the
"Effective Field Capacity."

	Portneuf silt loam		Ritzville loam	
	0-60 cm	0-105 cm	0-60 cm	0-105 cm
W_o, cm	21.4	38.8	19.8	41.5
m	0.043	0.043	0.106	0.111
Daily E_t, cm	.8	.8	.8	.8
W_D, cm	.49	1.61	2.11	7.30
$W_o - W_D$, cm	20.91	37.19	17.69	34.20
t, days	1.7	2.7	2.9	5.7

G. Modifications Underway

If the water table level is relatively close to the
soil surface or close to the bottom of the root zone, an
adjustment can be included in the crop coefficients to account for the movement of water upward from the saturated
zone using concepts presented by Gardner (1958). For bare
soil, the amount of water moving upward, W_u, can be estimated using:

$$W_u = 0.9 \left(\frac{z_c}{z_w} \right)^n E_{tp} \tag{20}$$

$$\text{for } z_w \geq z_c$$

with a crop,

$$W_u = \left[1 - \frac{AM - a_4}{100 - a_4} \right] \left(\frac{z_c}{z_w - z_r} \right)^n E_t \tag{21}$$

where z_c is the effective height of the capillary fringe
above the water table, cm; z_w is the depth to the water table, cm; z_r is the depth of roots, cm; AM is available soil

water to z_r in percent; a_4 is a constant (about 25); and n
is a constant for a given soil and crop (expected to vary
between 1 and 3). An example of this approximation of W_u
is presented in Figure 1 using unpublished data from L.N.
Namken (USDA-ARS-SWC, Weslaco, Texas). (Constants used in
equation (20) are z_w = 273 cm, z_c = 50 cm, n = 1.22, and
a_4 = 25%.)

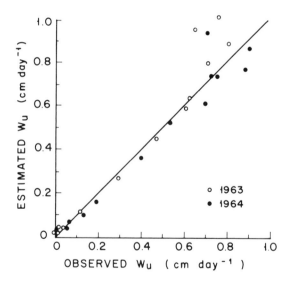

Figure 1. Comparison of estimated weekly mean amount
of water moving upward, W_u, with observed W_u. (Observed
data provided by L.N. Namken, USDA-ARS-SWC, Weslaco, Texas.)

This adjustment will require additional information on the expected rooting depth and the effective capillary fringe for the soil involved. It will also require current information on the existing depth to the water table as part of the input data.

An additional, optional subroutine is being developed to predict optimum timing of limited irrigations for water-short areas, or where irrigation water is expensive. Each time the program is run, it will estimate the soil water depletion throughout the balance of the season and the probable yield reduction if no irrigations are given. It will then predict the optimum time for applying specific increments of water under expected climatic conditions. A functional relationship between the effects of limited water and yields, which is difficult to define at this time for most crops, will be required.

H. Future Improvements Needed

The present computer program can provide acceptable estimates of soil water depletion and dates of future irrigations. However, a number of components need to be refined or added. Some of these are listed below:

1. Plant growth models are needed which are based on microclimate, plant characteristics, plant densities, soil water, and nutrients, and which can be easily calibrated.
2. A better procedure is needed for estimating evaporation from the soil surface. The procedure should allow simple calibration for various soils. When combined with effective leaf resistance to the diffusion of water vapor obtainable from item 1, this procedure will increase the accuracy of K_c in equation (8).
3. Better estimates of soil heat flux G are needed in equations (1) and (7).
4. Standardized meteorological data sites and meteorological measurements specifically for this use will improve estimates of E_t for the reference crop.
5. Yield-soil water level functions for all stages of growth and levels of evaporation are needed to permit predicting the adverse effects of delayed irrigations and optimum timing of limited irrigation water.
6. Distribution of irrigation water applied, especially for surface irrigation systems, would increase the

accuracy of all projected irrigations.

Summary

Irrigation programming practices have not changed appreciably in many areas during the past three decades because suggested techniques have not been acceptable, and the direct and indirect effects of excessive water use were not readily apparent. Several alternative methods of programming irrigations are now available. The scheduling of irrigations for each field using meteorological, soil, and crop data, coupled with field inspection, appears to be an economical and acceptable irrigation management service in the U.S.A. An irrigation management service requires technical competence and a good communications network. Reliable meteorological data also require technical competence, and periodic calibration of instruments.

The USDA-ARS-SWC irrigation scheduling computer program using meteorological techniques and soil-crop data is described in this chapter. Since its development and use for several years in Arizona, Idaho, and Nebraska, several modifications have been completed and others are under way. This problem has enabled private firms and service agencies to gain experience in providing irrigation management service while additional refinements are under way.

References

1. Baier, W. (1967). Recent advancements in the use of standard climatic data for estimating soil moisture. Ann. Arid Zone 6, 1-21.
2. Baier, W. (1969). Concepts of soils moisture availability and their effect on soil moisture estimates from a meteorological budget. Agr. Meteorol. 6, 165-178.
3. Baver, L.D. (1954). The meteorological approach to irrigation control. Hawaiian Planter's Record 54, 291-298.
4. Brown, R.J. and Buchheim, J.F. (1971). Water scheduling in southern Idaho, "A Progress Report," USDI, Bur. of Reclamation. (Presented at the Nat. Conf. on Water Resources Eng., ASCE, Phoenix, Ariz., 1971.)
5. Budyko, I.I. (1956). The heat balance of the earth's surface. U.S. Dept. Com. Weather Bur. PB 131692 (translated by N.A. Stepanova, 1958).

OPTIMIZING THE SOIL ENVIRONMENT

6. Corey, F.C. (1970). Irrigation scheduling by computer. Presented at the Joint Conv. of the Neb. State Irrig. Assn. and the Neb. Reclamation Assn., Grand Island, Neb., 1970.
7. Das, U.K. (1936). A suggested scheme of irrigation control using the day-degree system. Hawaiian Planter's Record 37, 109-111.
8. Denmead, O.T. and Shaw, R.H. (1959). Evapotranspiration in relation to the development of the corn crop. Agron J. 51, 725-726.
9. Fitzpatrick, E.A. and Cossens, G.G. (1966). Applications of Penman's and Thornthwaite's methods of estimating transpiration rates to determination of moisture of three Central Otago soils. N.Z. J. Agric. Res. 9, 985-994.
10. Fitzpatrick, E.A. and Stern, W.R. (1965). Components of the radiation balance of irrigated plots in a dry monsoonal environment. J. Applied Meteorol. 4, 649-660.
11. Franzoy, C.E. and Tankersley, E.L. (1970). Predicting irrigations from climatic data and soil parameters. Trans. Amer. Soc. Agr. Engs. 13, 814-816.
12. Fritz, S. (1949). Solar radiation during cloudless days. Heating & Ventilating 46, 69-74.
13. Gardner, W.R. (1958). Some steady state solutions of the unsaturated moisture flow equation with application to evaporation from a water table. Soil Sci. 85, 228-232.
14. Goss, J.R. and Brooks, F.A. (1956). Constants for empirical expressions for downcoming atmospheric radiation under cloudless skies. J. Meteorol. 13, 482-488.
15. Heermann, D.F. and Jensen, M.E. (1970). Adapting meteorological approaches in irrigation scheduling. Proc. ASAE Nat. Irrig. Symp., Lincoln, Neb., 1970, 00-1 to 00-10.
16. Israelsen, O.W., et al. (1944). Water application efficiencies in irrigation. Utah Agr. Exp. Sta. Bul. 311.
17. Jensen, M.E. (1968). Water consumption by agricultural plants. In "Water Deficits and Plant Growth" (T.T. Kozlowski, ed.), Vol. II, pp. 1-22. Academic Press, New York.
18. Jensen, M.E. (1969). Scheduling irrigations using computers. J. Soil and Water Conserv. 24, 193-195.
19. Jensen, M.E., Robb, D.C.N., and Franzoy, C.E. (1970). Scheduling irrigations using climate-crop-soil data.

20. Jensen, M.E., Wright, J.L., and Pratt, B.J. (1971).
Estimating soil moisture depletion from climate, crop,
and soil data. Trans. Amer. Soc. Agr. Eng. Winter
Meeting, 1969, Paper No. 69-641 (in press).

21. Jensen, M.E. and Heermann, D.F. (1970). Meteorological
approaches to irrigation scheduling. Proc. Amer. Soc.
Agr. Eng. Nat. Irrig. Symp., Lincoln, Neb, 1970, NN-1
to NN-10.

22. Makkink, G.F. and van Hermst, H.D.J. (1956). The actual evapotranspiration as a function of the potential
evapo-transpiration and the soil moisture tension.
Neth. J. Agr. Sci. 4.

23. Marshall, W.G. (1971). Operating a farm consulting
business. Proc. Sprinkler Irrig. Tech. Conf., Denver,
Colo., 1971.

24. Miller, D.E. (1967). Available water in soil as influenced by extraction of soil water by plants. Agron. J.
59, 420-423.

25. Miller, D.E. and Aarstad, J.S. (1971). Available water
as related to evapotranspiration rates and deep drainage. Soil Sci. Soc. Amer. Proc. 35, 131-134.

26. Ogata, G. and Richards, L.A. (1957). Water content
changes following irrigation of bare-field soil that is
protected from evaporation. Soil Sci. Soc. Amer. Proc.
21, 355-356.

27. Penman, H.L. (1948). Natural evaporation from open water, bare soil and grass. Proc. Roy. Soc. A93, 120-145.

28. Penman, H.L. (1952). The physical basis of irrigation
control. Proc. Int. Hort. Cong., London 13, 913-924.

29. Penman, H.L. (1963). Vegetation and hydrology. Tech.
Commun. No. 53, Commonwealth Bur. of Soils, Harpenden,
England.

30. Pierce, L.T. (1958). Estimating seasonal and short-
term fluctuations in evapo-transpiration from meadow
crops. Bul. Amer. Meteor. Soc. 39, 73-78.

31. Pierce, L.T. (1960). A practical method of determining
evapotranspiration from temperature and rainfall.
Amer. Soc. Agr. Eng. Trans. 3, 77-81.

32. Pruitt, W.O. and Jensen, M.C. (1955). Determining when
to irrigate. Agr. Eng. 36, 389-393.

33. Rickard, D.S. (1957). A comparison between measured
and calculated soil moisture deficit. N.Z. J. Sci. and
Tech. 38, 1081-1090.

34. Ritchie, J.T. (1971). Dryland evaporative flux in a

subhumid climate: I. Micrometeorological influences. Agron. J. 63, 51-55.
35. Ritchie, J.T. and Burnett, E. (1971). Dryland evaporative flux in a subhumid climate: II. Plant influences. Agron. J. 63, 56-62.
36. Rosenberg, N.J. (1969). Seasonal patterns in evapotranspiration by irrigated alfalfa in the Central Great Plains. Agron. J. 61, 879-889.
37. Thornthwaite, C.W. (1948). An approach toward a rational classification of climate. Geograph. Rev. 38, 55-94.
38. Tyler, C.L., Corey, G.L., and Swarner, L.R. (1964). Evaluating water use on a new irrigation project. Idaho Agr. Exp. Sta. Res. Bul. No. 62.
39. van Bavel, C.H.M. (1960). Use of climatic data in guiding water management on the farm. In "Water and Agriculture," pp. 80-100. Amer. Assn. Advan. Sci.
40. van Bavel, C.H.M. (1966). Potential evaporation: The combination concept and its experimental verification. Water Resources Res. 2, 455-467.
41. van Bavel, C.H.M. and Wilson, T.V. (1952). Evapotranspiration estimates as a criteria for determining time of irrigation. Agr. Eng. 33, 417-418, 420.
42. van Wijk, W.R. and de Vries, D.A. (1954). Evapotranspiration. Neth. J. Agr. Sci. 2, 105-119.
43. Wilcox, J.C. (1960). Rate of soil drainage following irrigation. II. Effects on determination of rate of consumptive use. Can. J. Soil Sci. 40, 15-27.
44. Willardson, L.S. (1967). "Irrigation Efficiency in the Escalante Valley, Utah." Utah Resources Series 37, Utah Agr. Exp. Sta.
45. Wright, J.L. and Jensen, M.E. (1971). Peak water requirements for southern Idaho. Proc. Amer. Soc. Civ. Eng., Irrig. and Drain. Div. (in press).

WATER UTILIZATION BY A DRYLAND ROWCROP

H.R. Gardner
Agricultural Research Service
U.S. Department of Agriculture

In the western United States, the word "dryland" carries the connotation that water is very often the limiting factor of crop production and that irrigation water will not be used to supplement the natural supply. The problem of increasing the efficiency of water use under dryland conditions can be approached two ways. These are: restricting losses of water directly from the soil or increasing efficiency of water use by the plants. A few aspects of the interaction between the two approaches will be examined in this paper.

Precipitation received at the soil surface is subject to loss by runoff, evaporation from the soil surface, and deep percolation. Runoff will not be considered here. The relative importance of both evaporation and deep percolation depends to some degree on the rainfall distribution and its relation to the growing season. At the Central Great Plains Field Station near Akron, Colorado, the major portion of precipitation falls in April, May, June, July, and August (89% in 1968). Sorghum is usually planted in the latter part of May and is harvested after frost about October 1. Thus the major portion of the high rainfall period overlaps the early part of the growing season.

As a test case let us examine sorghum production (Sorghum vulgare, var. Nebraska 505) at Akron, Colorado, in 1968. The sorghum was grown in rows spaced one meter apart with approximately 12 plants per meter within the row.

The cumulative precipitation and evapotranspiration for the season are presented in Figure 1. Precipitation had amounted to 9.7 cm by June 12 when the sorghum was planted, with 7.5 cm of this coming after April 1. One sprinkler irrigation of 4.5 cm water was added on June 23 to supplement the low soil moisture storage at the beginning of the season and is included in the precipitation. Irrigation is

not a normal dryland practice. The combination of the irrigation plus the natural rainfall for the season provided
sufficient water so that the evapotranspiration did not
surpass the water supply until the latter part of August.

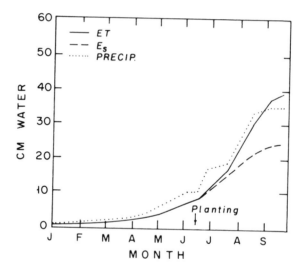

Figure 1. Cumulative precipitation, evaporation, and
evapotranspiration for 1968; sorghum with 1 meter row spacing.

The evaporation from the soil (E_s) is also shown in
Figure 1. The evaporation before planting was obtained
from a nearby bare lysimeter. The between-row evaporation
was computed using the relationship given by De Wit (1958):

$$P = m \, W/E_o \, ,$$

where P = dry matter production
       W = total transpiration used in the production of P
       E_o = daily evaporation rate
             (based on loss from free water surface)
       m = proportionality factor.

Using an m value of 210 kgm/ha day, approximately the
same as that computed by De Wit, a comparison of the evaporation component computed from the dry matter production
and evapotranspiration was made with the evaporation from
the bare lysimeter during the growing season. This comparison showed little difference between E_s values for the two
cases early in the growing season, indicating that the assumed value of m was adequate. Figure 2 shows the evaporation as a fraction of the total ET with time throughout the
growing season. This figure also shows the evaporation
fraction for an irrigated field that will be discussed
later. The fraction lost by evaporation for the entire
growing season was 0.53 for the dryland case. The last
dryland value from day 85 to day 102 is questionable, but
even if it were to be zero rather than 0.24 it would only
shift the fraction lost by evaporation during the entire
season from 0.53 to 0.51.

The evaporation rate from the sorghum plot during the
growing season dropped almost linearly from 100% at planting to 24% late in the season with a total loss of 17.7 cm
water. The cumulative evapotranspiration for the growing
season was 33.3 cm water.

Since the evaporation component of evapotranspiration
was calculated by difference between soil moisture loss and
transpiration, the evaporation term includes any deep
drainage that may have taken place. Deep drainage is defined as any water that has drained to a depth in the soil
below which no significant amount of water can be obtained
by the plant roots. Observations of water content profiles
throughout the growing season at Akron have indicated that
rooting is negligible below 120 cm.

If spring rains are assumed to contribute a significant
amount of storage water starting about April 1, and if
plant roots are assumed to reach full rooting depth by July
20, then any significant drainage should be during this 111
day period. There are two ways that drainage can be determined. The simplest method is to measure the water content
in the soil profile at the beginning and end of the period
of concern. The depth of measurement must be sufficient to

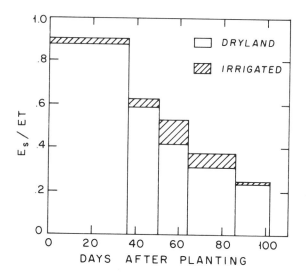

Figure 2. Evaporation as a fraction of evapotranspiration during the growing season for one meter row spacing sorghum for 1968 dryland and irrigated plots.

ensure that no water has moved below the measured zone. However, this method is not entirely satisfactory since water will move downward through the zone considered as long as there is a gradient in that direction even though the water content may not change. The measured water content change from 120 to 270 cm depth between April 1 and July 20 was approximately 0.5 cm which is of the same order of magnitude as the measurement error of the neutron method of determining water content.

The other method to evaluate water loss by downward movement is to determine the hydraulic conductivity for the

water content or suction range involved. This determination can be made by applying water to the soil surface,
covering the surface to stop evaporation, and calculating
the rate of flow by water content change with time. Then,
with the assumption of a unit gradient, it is possible to
calculate the hydraulic conductivity. The unit gradient
assumption is generally acceptable until the plant roots
approach the zone being considered.

The water content of the soil profile in 1968 at Akron
was low compared to that for a normal spring. Thus it was
expected that any loss by deep drainage would be quite
small.

In principle, deep movement of water below the root
zone can be significant if the hydraulic conductivity is
high enough. For example, conductivity at 0.3 bar suction
for the Rago soil at Akron is approximately 0.04 cm/day at
120 cm depth. Thus if the amount of soil water in the profile brought the suction to 0.3 bar by May 1, then the soil
could transmit about 2.5 cm water downward through that
zone assuming a unit gradient and 60 days time before the
plant roots could make a significant change in water content.

The contributions to the total water use by evaporation
and transpiration for the entire season starting from the
harvest of the previous crop are presented in Table 1.

The loss of water by evaporation from the previous harvest to December 31, 1967, exceeded the water application
by 2.7 cm or 66%. This loss, plus only 4.1 cm water being
added during a period of 107 days, is a good indication
that the soil surface was dry for all but a few hours of
the total time. The period January 1 to March 31 had 2.2
cm water added in 91 days with 1.6 cm lost. Thus 6.3 cm
were added and 8.4 cm were lost by evaporation in a time
period of 198 days. The next 72 days (April 1 to June 12)
had 7.5 cm of precipitation with 6.1 cm lost. The total
evaporative loss from the previous harvest to planting was
14.5 cm. The evaporative loss during the growing season to
September 22, the end of the data period, was 17.6 cm.
Thus the soil water loss during the growing season was 21%
greater than the loss from the previous harvest to planting. The average loss rate during the fallow period was
0.05 cm per day with 0.17 cm the average loss per day during the growing season. Thus it would appear that efforts
to conserve soil water should be aimed at the water lost

during the growing season. The normal technique for reducing evaporative loss is a mulch and it would be much more effective during periods when precipitation rates are high since the main benefit is the reduction of the constant-rate evaporation when the soil surface is wet (Greb et al., 1967).

TABLE 1
Precipitation, Evaporation, and Transpiration
for Dryland Sorghum at Akron, Colorado, 1968 Season.

| Time period | Precipitation | Loss | |
| | | Evaporation | Transpiration |
	cm	cm	cm
Sept. 15 to Dec. 31, 1967	4.1	6.8	
Jan. 1 to March 31, 1968	2.2	1.6	
April 1 to June 11, 1968	7.5	6.1	
Total for fallow period	13.8	14.5	
June 12 (planting) to Sept. 22, 1968	25.7	17.6	15.7
Total for year	39.5	32.1	15.7

If a conventional wheat straw mulch (which according to Gardner and Gardner [1969], can give about 15% reduction in evaporation during a fallow year) were applied to the field from which the foregoing data were taken, the water saving presented in Table 2 would presumably result. If a more efficient method of reducing loss, such as plastic sheeting, could reduce evaporation by a hypothetical 90%, then the total possible saving would be an increase in water available for transpiration of 185% as compared to 31% for the wheat straw mulch. These values assume the same efficiency for catching snow as the unmulched field and also assume that the field can be planted without destroying the mulches or their effectiveness.

TABLE 2
Calculated Water Saving from
Two Hypothetical Soil Mulches.

Time period	Water saved due to mulch	
	Mulch efficiency*	
	90%	15%
	cm	cm
September 15 to December 31, 1967	6.12	1.02
January 1 to June 12, 1968	6.93	1.15
Total for fallow period	13.05	2.17
June 12 to September 22, 1968	15.84	2.64
Total saved	28.89	4.81
Percent increase in water for transpiration	185%	31%

*Reduction of water loss rate

When a mulch is placed on the surface of the soil and evaporation is reduced, there is excess energy remaining that was used to evaporate water in the no mulch case. Some of this energy becomes available for transfer by advection and radiation to the plant leaf surfaces.

If the relationships discussed by De Wit are taken literally in the strictest sense and if it is assumed that all energy that was previously used by evaporation of soil water is now used by the crop, then the free water evaporation rate, E_o, in his equation will be increased in the same proportion as the transpiration, W, is increased. Then the ratio W/E_o would not change and the dry matter production would not increase. Obviously this is not the case in the field, but if any of the excess energy is used in this manner it would reduce the water use efficiency and make it more difficult to predict the production-transpiration relationship for that field.

The sorghum experiment discussed above had rather low seasonal rainfall and low water storage at the beginning of the crop season. The same year at Akron other sorghum plots were sprinkler irrigated with approximately 5 cm of water each time the soil water suction dropped to 1 bar at

the 30 cm depth. This provided adequate water for the entire growing season. Two irrigations were applied before
planting in April and May to fill the profile but no water
content measurements were made until planting time.

The total water loss to evaporation and deep drainage
was calculated as 23.1 cm water and the evapotranspiration
was 43.6 cm. The water loss was 53% of the total or very
nearly the same as for the dryland plot in spite of the
fact that the evapotranspiration was 10.3 cm higher for the
irrigated plots.

The water loss by drainage during the growing season
was 4.75 cm as indicated by the increase in water content
of the soil from 120 to 270 cm from June 12 to September 22.
This leaves 18.35 cm water to be assigned to evaporation
from the surface or 42% of the total evapotranspiration.

The high water loss by drainage makes it worth-while to
examine the use of a barrier within the soil to restrict
the deep drainage. In the irrigated case 4.25 cm could be
saved if a 90% effective barrier were placed at 120 cm
depth within the soil. The value of a barrier within the
soil to restrict drainage would be questionable for dryland
since, when the rainfall is low enough for the added water
to be needed, the deep drainage is reduced sufficiently to
make the barrier unnecessary.

In summary, there are several points of this discussion
that appear to have some significance. Considerable water
can be lost during the time between the previous harvest
and planting time but the possibility of saving this with a
traditional mulch is small since the soil surface is dry a
very high percentage of the time. The water loss during
the early part of the growing season can be greater than
the loss during the fallow period. The chance of saving a
portion of this water is greater than that from the fallow
period since the soil surface is wet a greater part of the
time.

The possibility of a detrimental effect of advective
energy released by reduction of direct soil moisture evaporation and used by the plant needs to be examined to see if
it is significant. Deep drainage can be significant in
amount in wet years but the value of corrective measures
would be questionable in most seasons.

One of the major contributions that we can make in dryland agriculture is to bring about an understanding of the
factors relating crop production and water use so that we

may avoid or at least be ready for the inconsistencies of the weather and water supply.

References

1. De Wit, C.T. (1958). Transpiration and crop yields. Versl. Landbovwk. Onderz. No. 646, p. 88.
2. Gardner, H.R. and Gardner, W.R. (1969). Relation of water application to evaporation and storage of soil water. Soil Sci. Soc. Amer. Proc. 33, 192-196.
3. Greb, B.W., Smika, D.E., and Black, A.L. (1967). Effect of straw mulch rates on soil water storage during summer fallow in the Great Plains. Soil Sci. Soc. Amer. Proc. 31, 556-559.

THE CONTROL OF THE RADIATION CLIMATE OF PLANT COMMUNITIES

M. Fuchs

Volcani Institute of Agricultural Research, Israel

Introduction

The importance of the radiation climate of crops derives from its bearing on the heat balance and on the photosynthetic activity of green plants. However, in contrast with the great efforts invested in the physical and chemical improvement of soil environments, it seems that insufficient attention has been devoted to the controlled transformation of the radiation climate.

Manipulation of the foliage cover density which modifies the radiation absorption has been used for centuries by foresters to promote forest regrowth and to improve the shape of trees. It is also known that one of the beneficial effects of pruning operations in orchards is the enhancement of light penetration into the canopy. However, most of the techniques used to date are based on traditional and empirical knowledge rather than on physical theory.

The importance of light intensity in photosynthesis has prompted a large number of investigations of the radiation distribution inside canopies. The classical work of Monsi and Saeki (1953) describes the light intensity distribution of grass communities in the form of Bouguer's exponential law where the path light is measured in terms of cumulative leaf area index and the extinction coefficient in terms of the reciprocal of leaf area index. As measurements showed that the extinction coefficients were different for direct and diffuse solar radiation, and furthermore varied according to solar elevation (Anderson, 1964), extinction coefficient theory has been completed to include the angular relationship between the light beam and elements of the foliage (Isobe, 1962; de Wit, 1965; Chartier, 1966; Anderson, 1966; Cowan, 1968). Light distribution in model canopies has also been analyzed numerically by de Wit (1965).

The mathematical complexity of the resulting geometrical

problems limits the application of extinction coefficient
theory to highly idealized canopy structures which are seldom encountered in real crops. A *fortiori* the theory cannot determine the canopy structure that will produce a pre-
scribed radiation profile. The numerical approach of
de Wit for model canopies with specified foliage orientation and distribution is capable of defining the interaction between crop structure and the radiation profile. Unfortunately the method is rather tantalizing, as it requires tedious measurements of leaf orientation and distribution functions.

Simple extinction coefficient theories consider radiation transmission as if it occurs through gaps in the foliage only, and thus neglects the scattering properties of
the foliage elements. Discrepancies between these theories
and experimental data are accentuated by the fact that solar radiation is usually measured in broad spectral bands
which present considerable variation of the spectral reflectivity and transmissivity of leaves. The few general
treatments of this additional difficulty (Cowan, 1968;
Isobe, 1969) refer to idealized canopy models and these do
not provide directives for radiation balance control by
modification of the canopy structure. As a result of the
theoretical and technical shortcomings, the radiation balance control based upon crop architecture still belongs to
the "trial and failure" approach.

Considering that the scattering properties of foliage
elements affect the radiation balance of canopies, modification of reflectivity and transmissivity of leaves offers
a possible means of controlling the radiation climate.
White kaolinite coatings applied on leaves have been found
to increase their reflectivity and reduce their transmissivity (Abou-Khaled et al., 1970). Reduction of the radiation load by the reflective coating reduced the transpiration rate of single leaves by more than 20%. Photosynthesis was decreased by the treatment when the incident radiant flux density was below 0.6 cal cm^{-2} min^{-1}, but was increased at higher flux densities. In all cases the transpiration ratios were reduced. However, generalization of
these findings on single leaves to complete crop covers is
uncertain because the canopy geometry modifies the pattern
of radiation absorption and the processes of water vapor
transport.

Preliminary results of canopy and soil coating

experiments by Seginer (1969), using lime solutions of various concentrations in a cotton field, do not show significant reduction of water use although the spraying increased the solar reflectance of the fully developed cotton crop from 0.18 to 0.24. Whitening the soil underneath the foliage alone increased the reflectance of the crop to 0.20. Spreading of magnesium carbonate on a bare loessial soil increased the solar reflectance from 0.31 to 0.64 immediately after application, and to 0.40 three months later (Stanhill, 1965). Evaporation from the treated plots was reduced by 20% and maximum temperatures near the soil surface were depressed by 10°C. Similar soil temperature and evaporation reductions are reported for soils coated with white petroleum resin emulsions (Gerard and Chambers, 1967)

If the crop canopy transmits solar radiation, changing the reflectivity of the soil surface underneath the foliage will modify the radiation level in the canopy. Pendleton et al. (1966) reported larger corn yields on white plastic covered soil than on soil with a black plastic cover and assigned the effect to the increase of light intensity in the lower part of the canopy. Light enrichment of corn rows by sloping aluminum reflectors resulted in a 26% increase of grain yield (Pendleton et al., 1967). In contrast to these findings, Aase et al. (1968) reported higher corn grain yields on black coated soil than on white coated soil. This result contradicts what is known about the photosynthesis characteristics of corn. Modification of the solar radiation balance by the black coating apparently increased soil temperature and thus improved the growing conditions of the seedlings.

Addition of artificial light in the top, middle and bottom part of the canopy of field grown soybean plants has been found to increase the productivity of the lower portion of the canopy (Johnston et al., 1969). This last result emphasizes the importance of light distribution within the canopy and the unresolved problem of the interaction between canopy geometry and the radiation climate. On the other hand the promising results obtained by soil surface coatings on both evaporation reduction and photosynthesis enhancement cannot be generalized and translated into rational agricultural treatment without establishing their effect on the radiation balance of crop canopies. A better understanding of the interactions between soil and foliage optical properties is clearly needed. The aim of this paper is to present an approach to this problem which avoids the theoretical and technical difficulties of the extinction coefficient theories.

The Solar Radiation Balance in a Vegetation Stand

The interaction between solar radiation and the vegetation stand (canopy and soil) can be described in macroscopic terms by considering the three accessible boundaries of the system: the top of the canopy, the bottom of the canopy, and the soil surface. We assume that these boundaries occupy distinct locations in space, restricting the approach to canopies which do not lie on the ground. Having defined the three geometrical boundaries of the system, we can place a set of four suitable radiation sensors to measure the solar radiation flux densities reaching and leaving these boundaries, and thus determine the solar radiation balance of the canopy and the soil separately. The relationship between the four measured flux densities depends upon the optical and geometrical properties of the soil-canopy system. In this system the soil is considered opaque.

As the direct and diffuse components of the solar radiation do not follow the same optical path through the canopy, we shall consider them separately.

A. Diffuse Solar Radiation

Consider a vegetation stand irradiated by a uniformly diffusing hemisphere. Above the canopy, D_1 is the downward diffuse radiant flux density, and D_2 is the upward flux density. Underneath the canopy, D_3 represents the flux density directed toward the soil, and D_4 the flux density reflected by the soil (Figure 1).

The set of equations which describes the relationship between D_1, D_2, D_3, and D_4, is:

$$D_2 = \rho_\ell x D_1 + \tau_\ell x D_4 + (1 - x)D_4 \tag{1a}$$

$$D_3 = \rho_\ell x D_4 + \tau_\ell x D_1 + (1 - x)D_1 \tag{1b}$$

$$D_4 = \rho_s D_3 \tag{1c}$$

where x, the canopy cover density is the probability for

diffuse solar radiation impinging on a boundary of the canopy to encounter an element of the canopy. Rescattering of the radiation depends upon ρ_ℓ and τ_ℓ, the reflectance and transmittance of the canopy, and on ρ_s the reflectance of the soil.

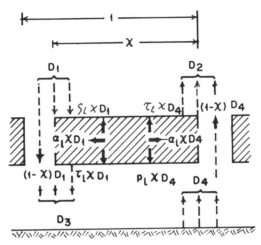

Figure 1. Components of the solar radiation balance in a vegetation stand irradiated by diffuse solar radiation.

The optical characteristics of the system, ρ_ℓ, τ_ℓ and ρ_s vary with wavelength in the solar spectrum. Consequently (1) is valid only if D_1 is restricted to a narrow spectral waveband. As the most significant change in the optical properties of foliage occurs at a wavelength of 700 nm, this condition is satisfied by separating the solar spectrum into visible and infrared radiation.

B. Direct Solar Radiation

Direct solar radiation, R_1, impinging on the vegetation stand can be scattered either by the soil or by the canopy. Direct solar radiation reaching the soil is reflected in proportion to $(1 - p)\rho_s$, where p is the probability for the solar beam to encounter an element of the canopy and ρ_s is the reflectance of the soil.

The scattering of direct solar radiation by the canopy is proportional to $(\rho_\ell + \tau_\ell)p$, where ρ_ℓ and τ_ℓ are the reflectance and the transmittance of the canopy, respectively.

Let R_2 be the upward flux density at the upper boundary of the canopy, R_3 the downward flux density at the bottom of the canopy, and R_4 the flux reflected by the soil (Figure 2). The incident flux density R_1 equals R cos i, where R is the radiant flux density impinging on a surface normal to the beam, and i is the angle of incidence.

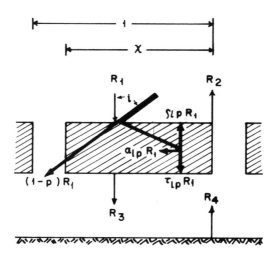

Figure 2. Components of the solar radiation balance in a vegetation stand irradiated by beam solar radiation.

Assuming that the radiation scattered by the plants and the soil is completely diffuse, and that the values of ρ_s, ρ_ℓ and τ_ℓ are independent of the radiation path in the canopy, the following relationships between R_1, R_2, R_3 and R_4 exist:

$$R_2 = \rho_\ell p R_1 + x \tau_\ell R_4 + (1 - x) R_4 \tag{2a}$$

$$R_3 = \tau_\ell p R_1 + x \rho_\ell R_4 + (1 - p) R_1 \tag{2b}$$

$$R_4 = \rho_s R_3 \tag{2c}$$

The ground cover x is related to p by:

$$x = \frac{1}{2\pi} \int_0^{2\pi} \int_0^{\pi/2} p \cos i \, di \, dn \tag{3}$$

where n is the azimuthal angle.

The value of p in (2) is a function of the path fol-
lowed by the beam in the canopy. Consequently it is a
function of the solar azimuth and the solar elevation. A
direct determination of p can be obtained either by monitoring sunfleck frequencies or by scanning hemispherical
photographs of the sky from the soil surface through the
foliage. The analysis of a series of such photographs
taken in a fully grown wheat crop is summarized in Figure 3.
The azimuthal symmetry permits describing p as a function
of the angle of incidence only. The resulting numerical
integration of (3) yields a value of 0.40 for x, the ground
cover density.

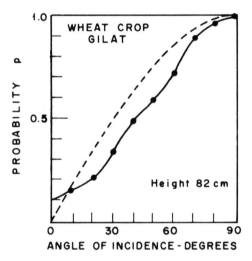

Figure 3. Angular variation of the interception probability by a canopy element of beam radiation.

As in the diffuse case, equations (2) are strictly valid only for monochromatic radiation. The division of the
solar spectrum at 700 nm usually satisfies this condition.
The assumption that τ_{ℓ} and ρ_{ℓ} have identical values for direct and diffuse radiation and that they are independent of
the radiation path in the canopy introduces a more serious
restriction. An analysis of visible and infrared reflection measurements (Fuchs et al., submitted) shows that this
assumption is acceptable in the case of a wheat canopy.

Control of the Diffuse Radiation Climate

It is convenient to characterize the radiation balance of vegetation stands using dimensionless flux densities normalized with respect to D_1, the incident diffuse radiation.

$$RD = D_2/D_1 \qquad (4a)$$

$$TD = (D_3 - D_4)/D_1 \qquad (4b)$$

$$AD = [(D_1 - D_2) - (D_3 - D_4)]/D_1 \qquad (4c)$$

Since the vegetation stand as a whole constitutes an opaque system, its diffuse solar radiation balance depends only upon RD, which defines the reflection coefficient of the vegetation stand. If RD increases, the solar radiation absorption by the system lessens. The partition of the solar radiation in the system between the soil and the canopy is governed by TD, the effective transmission of the canopy, and by AD, the effective absorption by the canopy. The energy conservation law requires that:

$$RD + TD + AD = 1 \qquad (5)$$

We define:

$$r = x\rho_\ell \qquad (6a)$$

$$t = 1 - x(1 - \tau_\ell) \qquad (6b)$$

$$a = x\alpha_\ell \qquad (6c)$$

where α_ℓ is the absorptance of the canopy alone. By definition we have:

$$\alpha_\ell + \rho_\ell + \tau_\ell = 1 \qquad (7)$$

and consequently

$$a + r + t = 1 \qquad (8)$$

The parameters a, r, and t combine the optical and geometrical characteristics of a plant canopy irradiated by diffuse radiation, and define the specific scattering properties of a given canopy.

The coefficient defined by equation (4) can be rewritten as functions of a, r, t and ρ_s:

$$RD = r + \rho_s t^2/(1 - r\rho_s) \qquad (9a)$$

$$TD = t(1 - \rho_s)/(1 - r\rho_s) \qquad (9b)$$

$$AD = a[1 + \rho_s(t - r)]/(1 - r\rho_s) \qquad (9c)$$

Equation (9) provides the basis for predicting the alteration of the diffuse solar radiation climate in a vegetation stand, resulting from changes of the scattering characteristics of the foliage and modification of the reflectance of the soil surface.

According to (9) the diffuse solar radiation balances of the canopy, of the soil and of the complete stand are hyperbolic functions of soil reflectance. The consequence is that control of the diffuse solar radiation climate by modification of the soil reflectance is the most effective at high values of this parameter.

A plot of RD as given in (9a) illustrates the possible modification of the radiation balance of an entire vegetation stand (Figure 4). In plant communities characterized by large values of x and consequently low values of t, the reflectance of the canopy is the dominant parameter. In this case, the most effective means of diminishing the absorption of solar radiation will be by reflective coatings of leaves, while the contribution of soil reflectance will be minor. It is however worthwhile noting that the absolute effect of an increase of soil reflectance will be larger for highly reflective foliage. Consequently reflective coatings of foliage and of soil in dense vegetation stands can produce a synergetic reduction of solar radiation absorption. Increasing the foliage reflectance of open vegetation stands like orchards, vineyards, etc., which have low values of x and consequently low values of r, should result in a minor decrease of the solar radiation absorption. In this case the most significant effects are likely to be obtained by reflective coating of the soil surface.

The absorption of diffuse solar radiation by the soil is controlled by TD. Equation (9b) shows that this coefficient is directly proportional to t, in agreement with common-sense evidence that the radiative load on the soil underneath a canopy increases when the interception of radiation by the canopy decreases. Absorption by the soil also

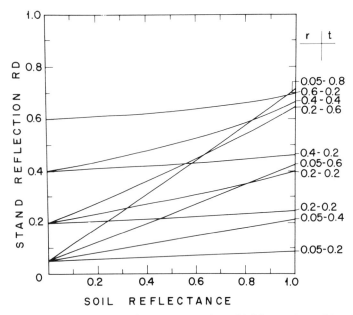

Figure 4. Stand reflection under diffuse irradiation.

depends upon the soil reflectance. The attenuation of effective radiation absorption by the soil surface resulting
from an increase of the soil reflectance is illustrated in
Figure 5. For low values of r (i.e., when only a small
fraction of the radiation is reflected by the canopy, either because the foliage is dark or because it is sparse)
the reduction of radiation absorption is approximately proportional to soil reflectance. At a fixed value of soil
reflectance, the relative absorption of radiation by the
soil should increase with r. Consequently, reflective
coatings of the foliage will increase the radiation absorption by the soil. This effect will be maximal at a soil
reflectance of about 0.6. A practical consequence is that
reflective coatings of foliage can be used to hasten the
warming of the root zone during spring and early summer.

The coefficient AD governs the diffuse solar radiation
balance of the foliage. According to (9c), radiation adsorption is directly proportional to a, which combines the
foliage absorptance and density. Maximum absorption will
occur when a = 1, but if any radiation reaches the soil
surface, the efficiency of the radiation absorption in the

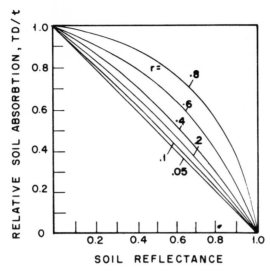

Figure 5. Relative change of radiation absorption by the soil underneath a canopy cover (diffuse radiation).

foliage, AD/a, is always enhanced by increasing the soil reflectance (Figure 6). The gain of radiation absorption is caused by a larger participation of the lower boundary of the canopy which is irradiated by the reflected flux density. Consequently the radiation distribution in the profile should be more uniform. However, as the theory omits deliberately the characteristics of the radiation profile in the canopy, it cannot provide quantitative information on this aspect of radiation balance control.

For a constant value of t, AD/a is an increasing function of r, but the absolute value of AD decreases with r. Large values of r increase the curvature of the hyperbole branches in Figure 6 and consequently accentuate the effect of soil reflectance change at high values of the soil reflectance. Figure 6 also shows that the increase of soil reflectance enhances absorption most effectively in canopies with large values of t.

The spectral reflectance of soil surface coatings can also modify the spectral composition of the radiation absorbed by foliage. High soil reflectance in the photosynthetic wavelengths will increase the absorption of photosynthetically active radiation. By contrast, lowering the infrared reflectance of the soil will reduce the radiative heat load on the canopy. Since the infrared reflectance of foliage is larger than the visible reflectance, changing the soil reflectance can result in a larger relative modification of the infrared radiation balance than of the visible radiation balance.

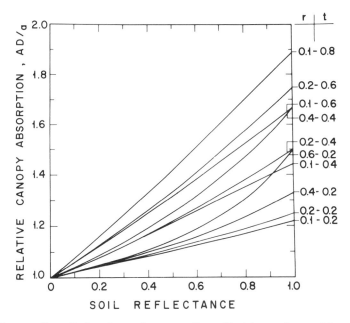

Figure 6. Relative change of radiation absorption by the canopy (diffuse radiation).

Control of the Radiation Climate for Beam Incidence

The theory of radiation balance control for beam incidence parallels the treatment for diffuse radiation. The analytical presentation is somewhat more complex because the angle of incidence of incoming radiation must be taken into consideration.

The dimensionless radiation fluxes normalized with respect to R_1 (which is the radiant flux density on a horizontal surface) are as follows:

$$RR = R_2/R_1 \tag{10a}$$

$$TR = (R_3 - R_4)/R_1 \tag{10b}$$

$$AR = [(R_1 - R_2) - (R_3 - R_4)]/R_1 \tag{10c}$$

where RR, TR, and AR are respectively the reflection coefficient of the entire vegetation stand, the effective absorption fraction of the soil, and the effective absorption fraction of the canopy. We have also:

$$RR + TR + AR = 1 \tag{11}$$

We combine the optical and geometrical properties of the foliage:

$$r' = p\rho_\ell \tag{12a}$$

$$t' = 1 - p(1 - \tau_\ell) \tag{12b}$$

$$a' = p\alpha_\ell \tag{12c}$$

and
$$r' + t' + a' = 1 \tag{13}$$

where r', t' and a' are, respectively, the fractions of direct radiation reflected, transmitted and absorbed by the canopy. These parameters are linear functions of the probability p, which depends upon the angle of incidence.

The balance equations in (2) permit expressing the fluxes defined in (10) as functions of the optical and geometrical characteristics of the plant community:

$$RR = r' + \rho_s tt'/(1 - r\rho_s) \tag{14a}$$

$$TR = t'(1 - \rho_s)/(1 - r\rho_s) \tag{14b}$$

$$AR = [a' + (at' - a'r)\rho_s]/(1 - r\rho_s) \tag{14c}$$

The form of (14) is identical to (9). Consequently control of the direct solar radiation balance of the vegetation stand, of the soil underneath the canopy, and of the foliage is governed by the same rules as those of the diffuse solar radiation. However, whereas in the case of diffuse radiation variation of the controlled parameters produced constant effects, here the result will depend upon the solar elevation, i.e. the hour of the day and the season of the year. The angular distribution of p establishes the variation of the dimensionless fluxes as a function of

the angle of incidence. This additional source of variations, introduced by p, also provides another means of controlling the solar radiation balance, as the relationship
between p and the angle of incidence depends upon canopy
architecture.

To illustrate the changes occurring in the fluxes given
by (14) when solar elevation varies, let us analyze an idealized example. We assume a canopy model for which p = sin
i. The integral in (3) yields x = 0.50. A common value
for both ρ_ℓ and τ_ℓ in the visible range of the solar spectrum is 0.10, which yield r = 0.05 and t = 0.55 according
to (6). Figure 7a shows that white soil coating can substantially increase the reflection of a vegetation stand.
However, the change of stand reflection varies with angle
of incidence. The reflection from the stand increases with
angle of incidence when ρ_s < 0.2, but decreases for larger
values of soil reflectance.

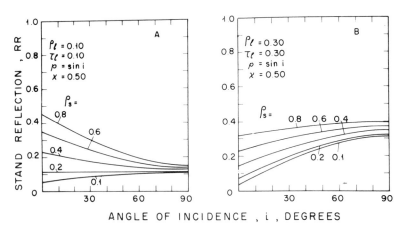

Figure 7. Stand reflection for the canopy model defined by p = sin i, under beam radiation.

In the near infrared solar spectrum ρ_ℓ and τ_ℓ are close
to 0.30, and thus r = 0.15 and t = 0.35 for the considered
canopy model. The resulting curves (Figure 7b) indicate
that stand reflection increases both with angle of incidence and with soil reflectance. The decrease noted in
Figure 7a cannot occur here because

$$d(RR)/d(\sin i) = 0.3 - 0.245\rho_s/(1 - 0.15\rho_s)$$

is always positive. The opposite trends for the angular variation of stand reflection in the visible and infrared ranges of the solar spectrum have been observed in a wheat crop, the geometry of which is close to that of the canopy model used in this example (Fuchs et al., submitted).

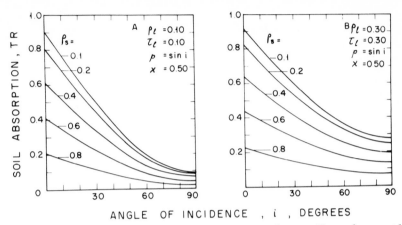

Figure 8. Radiation absorption of the soil underneath a canopy satisfying p = sin i (beam radiation).

According to (14b), at fixed angle of incidence the absorption of beam radiation by the soil underneath the canopy varies with soil reflectance in the same fashion as the absorption of diffuse radiation. The angular variations of the absorption imposed by the chosen canopy model are shown in Figures 8a and 8b. The absorption of radiation decreases linearly with the probability for a direct ray to reach the soil. However, larger values of soil reflectance attenuate the drop of absorption by the soil. Larger values of ρ_ℓ and τ_ℓ in Figure 8b, which increase absorption by the soil, also contribute to the attenuation of angular variation.

The absorption of direct solar radiation by the canopy is enhanced by increasing the reflectance of the soil (Figure 9). Maximum effects occur at high solar elevation, when the absolute value of incoming radiation is high. Thus, a favorable effect of radiation absorption increase due to soil whitening at a high latitude will not necessarily be reproduced at a lower latitude.

A comparison between parts a and b in Figure 9 indicates that for large values of ρ_ℓ and τ_ℓ the absolute effectiveness of a change of soil reflectance is reduced. The large ρ_ℓ and τ_ℓ also attenuate the angular variation of the canopy absorption. If we consider that Figures 9a and 9b characterize photosynthetic light and near infrared radiation, respectively, a spectrally uniform increase of soil reflectance should increase the proportion of photosynthetic light absorbed in the canopy. Furthermore it appears possible to modify the spectral composition of the radiation absorbed in the canopy by controlling the spectral reflectance of the soil coating. The range over which this modification can be operated, which is limited by the scattering properties of the canopy, can be assessed from (14c).

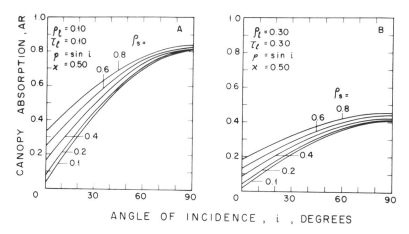

Figure 9. Radiation absorption by a canopy satisfying p = sin i (beam radiation).

The example shown in Figure 9 also points out that, where changes of the solar radiation balance modify critically the water balance or the soil temperature [as in the experiments reported by Aase et al. (1968) and Pendleton et al. (1966)] characterizing a coating as "white" or "black" with respect to visible radiation only does not describe adequately the soil treatment, since visible light constitutes less than 50% of the solar energy flux density. The reflectance measurements by Aase et al. (1968) contribute to the confusion because the silicon solar cells which were used have a maximum sensitivity in the near infrared spectral range and consequently responded to changes in either the flux density or the spectral composition.

Conclusions

Theoretical aspects of controlling the solar radiation balance in plant communities have been discussed on the basis of a macroscopic approach which neglects the detailed pattern of solar radiation absorption in the canopy profile. The vegetation stand is treated as a two-layer system in which the canopy top, canopy bottom, and opaque soil surface are the three limiting boundaries. The premises thus exclude the possibility of deducing any information on solar radiation distribution within the foliage. The resulting equations permit a quantitative assessment of the extent to which the radiation climate of the canopy, of the soil, and of the complete stand, can be controlled by changing density of plant cover, optical properties of foliage, and/or reflectance of soil.

Since in practice the modification of soil reflectance is the most easily controlled factor, its effect on the solar radiation balance has been thoroughly investigated. It appears that increasing the soil reflectance can reduce the absorption of solar radiation by the vegetation stand as a whole, but at the same time increase the absorption of solar radiation by the canopy. The largest changes in solar radiation balance can occur at high values of soil reflectance. The relative response to soil surface coating will therefore be largest for crops grown on light colored soils. The maximum effect of soil reflectance alteration is observed at high solar elevation.

The results of this analysis suggest which parameters ought to be investigated by rationally designed experiments aimed at modifying the radiation climate of plant communities. The controlling equations (9) and (14) are entirely defined if the probability p, the canopy cover density x, the scattering properties of the foliage τ_ℓ and ρ_ℓ, and the soil reflectance ρ_s, are known. Hemispherical photographs of the sky taken through the canopy from the soil surface provide a convenient technique for the determination of p and x. Direct measurements of ρ_ℓ and τ_ℓ are more difficult. An indirect method is offered by measuring the variation of the vegetation stand reflectance as a function of solar elevation. These data permit solving for τ_ℓ and ρ_ℓ the system of equations formed by (6), (12) and (14a). The

189

abrupt change of leaf optical properties near 700 nm makes
it necessary to carry out separate measurements for visible
and infrared radiation. As solar radiation always includes
a direct and a diffuse component, the true solar radiation
balance is represented by a weighted combination of (9) and
(14), which depends upon the ratio of direct to global solar radiation.

The components of the radiation balance underneath the
canopy, which are usually difficult to determine because of
sampling problems, are required to complete the analysis,
and to provide a check on the predictions by (9) and (14).
Independent measurements of the soil reflectance, either on
undisturbed soil sample or on the natural soil before crop
emergence, can also provide valuable information.

References

1. Aase, J.K., Kemper, W.D., and Danielson, R.E. (1968).
   Response of corn to white and black ground covers.
   Agron. J. 60, 234-236.
2. Abou-Khaled, A., Hagan, R.M., and Davenport, D.C.
   (1970). Effects of kaolinite as a reflective anti-
   transpirant on leaf temperature, transpiration, photosynthesis, and water-use efficiency. Water Resources
   Res. 6, 280-289.
3. Anderson, M.C. (1964). Light relations of terrestrial
   plant communities and their measurement. Biol. Rev.
   39, 425-486.
4. Anderson, M.C. (1966). Stand structure and light penetration. II. A theoretical analysis. J. Appl. Ecol.
   3, 41-54.
5. Chartier, P. (1966). Etude du microclimat lumineux
   dans la vegetation. Ann. Agron. 17, 571-602.
6. Cowan, I.R. (1968). The interception and absorption of
   radiation in plant stands. J. Appl. Ecol. 5, 367-379.
7. de Wit, C.T. (1965). Photosynthesis of leaf canopies.
   Agr. Res. Rept. 663. Centre for Agricultural Publications and Documentation, Wageningen.
8. Fuchs, M., Stanhill, G., and Waanders, A.G. Diurnal
   variations of the visible and near infrared reflectance
   of a wheat crop. Submitted for publication in Israel
   J. Agr. Res.
9. Gerard, C.J. and Chambers, G. (1967). Effect of reflective coatings on soil temperatures, soil moisture, and

the establishment of fall bell peppers. <u>Agron. J.</u> <u>59</u>, 293-296.

10. Isobe, S. (1962). Preliminary studies on physical properties of plant communities. <u>Bul. Nat. Inst. Agr. Sci.</u> (Japan) <u>A9</u>, 29-67.

11. Isobe, S. (1969). Theory of the light distribution and photosynthesis in canopies of randomly dispersed foliage area. <u>Bul. Nat. Inst. Agr. Sci.</u> (Japan) <u>A16</u>, 1-25.

12. Johnston, J.T., Pendleton, J.W., Peters, D.B., and Hicks, D.R. (1969). Influence of supplemental light on apparent photosynthesis, yield, and yield components of soybean (<u>Glycine</u> max L.). <u>Crop Sci.</u> <u>9</u>, 577-581.

13. Monsi, M. and Saeki, T. (1953). Über den Lichtfaktor in den Pflanzengesellschaften und seine Bedeutung für die Stoffproduktion. <u>Jap. J. Bot.</u> <u>14</u>, 22-52.

14. Pendleton, J.W., Peters, D.B., and Peek, J.W. (1966). Role of reflected light in the corn ecosystem. <u>Agron. J.</u> <u>58</u>, 73-74.

15. Pendleton, J.W., Egli, D.B., and Peters, D.B. (1967). Response of <u>Zea</u> <u>mays</u> L. to a "light rich" field environment. <u>Agron. J.</u> <u>59</u>, 395-397.

16. Seginer, I. (1969). The influence of liming on the increase of albedo of a cotton field. Progress Rept., Technion, Israel Inst. of Tech. Faculty of Agr. Eng., Haifa (in Hebrew).

17. Stanhill, G. (1965). Observations on the reduction of soil temperature. <u>Agr. Meteorol.</u> <u>2</u>, 197-203.

NUTRIENT SUPPLY AND UPTAKE IN RELATION TO
SOIL PHYSICAL CONDITIONS

R.E. Danielson
Colorado State University, U.S.A.

Introduction

The soil volume explored by the root system of a field crop is the reservoir supplying all of the chemical elements required for the growth of that crop with the exception of the carbon and oxygen obtained from the foliar atmosphere. Crop growth requires, among many processes, absorption of mineral nutrients from the soil. The absorption requires in turn that the nutrients be present in the soil, that they appear in solution in the proper chemical state, that they be available at the root-solution interface, and that the plant root be capable of absorbing them. Adequate growth (an admittedly complex concept) requires that the nutrients be absorbed by the plant in proper quantity and ratio to each other, and at a sufficient rate during the period of growth. Thus, adequate plant nutrition is possible only if each essential nutrient is present in the soil solution at the root surface in adequate quantity throughout the growing season and if the many other environmental and physiological factors are favorable for nutrient uptake, translocation, and utilization by the plant.

This discussion is concerned with the relation of the physical properties of the soil to requirements for adequate plant nutrition. The entire nutritional process might be simplified, in a convenient manner, by the process equation (1) where four nutrient conditions are assumed and three steps are indicated as requirements for transfer from the chemical supply reservoir to the point of utilization in the plant:

$$\text{Nutrient supply} \rightarrow \text{Available forms} \rightarrow \text{Contact with root surfaces} \rightarrow \text{Uptake and use by plants} \quad (1)$$

It has been generally recognized (Shaw, 1952) that the

physical or mechanical nature of the soil affects the
growth of an established plant through its influence on water, aeration, temperature, and the resistance offered to
root elongation and enlargement. It appears reasonable to
assume that the same four soil features also affect plant
nutrition. In addition, the nature of the solid phase in
terms of chemical composition, specific surface, and surface charge density has a direct influence on the supply of
some nutrients in the soil, their release to the soil solution, and their transport to the root.

The ultimate aim of a dissertation regarding the influence of soil physical conditions on plant nutrition should
be the quantitative evaluation of the physical influences
on the conditions and steps of equation (1). Furthermore,
the title to this panel requires that we elucidate the manner by which the agriculturist could effectively maintain
or alter the soil profile condition so as to improve soil
fertility and plant nutrition, or to improve the efficiency
of supplemental fertilization. Quantitative evaluations,
however, are not yet available for all facets of the system.
As is generally the case in soil-plant relations, the effects of soil physical conditions on nutrition are often
complex, interrelated, and indirect. Consideration at this
time must be limited to some more-or-less direct effects
which have been observed or which may be predicted by the
use of reasonable concepts.

Reviews have been written concerning the influence of
various soil physical conditions on plant activity and
growth. Page and Bodman (1951) reviewed the literature of
twenty years ago in relation to the effect of soil physical
properties on nutrient availability. At the same time,
Wadleigh and Richards (1951) considered soil water as it
relates to mineral nutrition of plants. The monograph edited by Shaw (1952) entitled "Soil Physical Conditions and
Plant Growth" has been a classic to soil edaphologists
since its publication. Wiersma (1959) and Pearson (1966)
have reviewed the influence of soil environment on root development, and Rosenberg (1964) has considered the response
of plants to the physical effects of soil compaction. In
each of these reviews, at least some attention has been
paid to the availability and uptake of nutrients. Perhaps
the most complete coverage of the soil-nutrient-plant system was given by Fried and Broeshart (1967). The more biochemical aspects of plant nutrition, including the

absorption and translocation process, have been covered in considerable detail by Steward and Sutcliff (1959) and by Bould and Hewitt (1963).

Let us direct our attention to the processes indicated in equation (1) and consider them in connection with the physical condition of the soil and its effect on plant nutrition.

Nutrient Supply

The soil matrix, including both mineral and organic components, functions as the storehouse for nutrients in the soil. The insoluble materials of the matrix do not readily release the elements of which they are composed, but may be associated with an important quantity of adsorbed ions which can be exchanged reversibly with those in solution. Salts of widely varying solubility comprise a third source of nutrient supply and include commercial fertilizers applied to the soil.

The supply of nutrients in the soil at a specific time takes different forms for the various nutrients. Most of the potassium and phosphorus is in mineral form and very little of the phosphorus is in solution. A great proportion of the nitrogen is in organic matter while most of the chloride and, in humid regions, much of the sulfate may be in solution.

If we may reason that alteration of the soil physical factors does not alter the native soil fertility expressed by the composition of the soil matrix and does not alter the application of fertilizers by the agriculturist, then our primary concern about supply is directed to the single nutrient, nitrogen. It must be recognized, of course, that the soil physical condition may, in fact, influence native nutrient supply through its relation to transport by wind or water erosion or to soil leaching.

Nitrogen supply in the soil is related to the processes of nonsymbiotic and symbiotic fixation by microorganisms. Hence, soil temperature, aeration, and water potential are influencing factors. It is recognized that nitrogen fixation, especially by legumes, may have real significance to agriculture in the tropics and perhaps in more temperate regions as well. However, the application of fertilizer nitrogen is becoming so commonplace in crop production, especially where agriculture is mechanized and high yields

195

are obtained, that the reliance on nitrogen fixation in the soil is greatly lessened. We may be justified then in neglecting further discussion of the role of soil physical factors in regard to the natural supply of nutrients in the soil.

Available Forms

In order to be of use in plant nutrition, nutrients in the soil must appear at the root surface in a form capable of being absorbed and utilized by the plant. This "available" form is a subject receiving considerable attention by soil chemists, and no simple definition of it may be accepted universally. However, for the present we shall assume that a nutrient is in an available form if it is in solution and in a chemical state such that uptake and beneficial utilization can take place. In most cases solubility and oxidation-reduction potential are the critical factors.

The soil physical condition probably plays its biggest role in conversion of nutrient supplies to available forms through its influence on organic matter mineralization. The release of organic nutrients, which is especially important in the case of nitrogen and many trace elements, is regulated by soil temperature, aeration, and water supply. Response of soil microorganisms to these variable factors is the important factor (McCalla, 1959). The release of nutrients from soil minerals can be expected to be influenced by the same factors. Soil aeration, through its relation to oxidation-reduction, may influence nutrient availability in differing ways. Reducing conditions may greatly enhance solubility of iron in the ferrous form and appreciably increase the soluble supply of phosphorus. Nitrification, however, may be restricted and toxic quantities of nitrite may accumulate.

Exchange of adsorbed ions with those in solution fluctuates reversibly with concentration of the solution, which in turn varies with soil water content as long as the solution is not saturated. Holliday (1970), in reporting many years of research at the University of Leeds, has concluded that crop response to irrigation appears to be related in many cases to increased availability of nutrients.

OPTIMIZING THE SOIL ENVIRONMENT

Nutrient Contact with Root Surfaces

The single condition of greatest importance in crop nutrition is the continuous renewal of nutrients at the location of root absorption. This condition requires either transport of the nutrients to the absorption sites or enlargement and elongation of the root system so that contact with the soil supply is maintained by the formation of new absorption sites. Both of these processes are important to crop production under field conditions. As pointed out by Fried and Broeshart (1967), a continuous description of the solution concentration at the root surface is necessary for the complete description of nutrient supply over the growing season.

Below critical limits, which vary widely among the different essential elements, the net uptake by plants is influenced by the nutrient concentration. This appears to be true whether the intake is primarily due to water entry, or diffusion, or metabolic activity. Data relating nutrient concentration to absorption are mostly from well-stirred solution culture studies. Discussion of such data may be found in the reports of Sutcliffe (1962), Wallace (1963), or Mitchell (1970). Influence of concentration of a particular nutrient on its accumulation in the plant varies with circumstances. The curve for phosphorus uptake rate by corn seedlings from a nutrient solution in Figure 1 is typical of many published results.

Soil physical conditions impose restrictions upon the maintenance of solute concentration at the root, through their influence on both root growth and nutrient movement in the soil solution. Their effects are apparent in Figure 1.

A. Root Extension and Proliferation

It is clear that extension of the crop root system into new soil volumes results in continued nutrient supply at new absorption sites. This is, however, not the only importance of root development. The enlargement of the rooting volume and increased proliferation decreases the total resistance to solute transfer from sources of supply to sites of absorption. This is particularly significant because root development is stimulated in soil regions where nutrients are locally more abundant (Fehrenbacher and

197

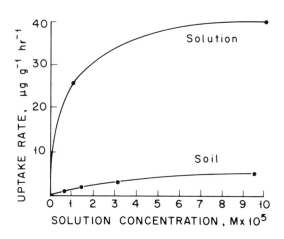

Figure 1.   Relation of solution concentration to rate
of nutrient absorption by corn seedlings from solution and
from a silty clay loam soil at 1/2 bar matric suction (after Watanabe et al., 1960).

Snider, 1954).
    The significance of root extension in helping to meet
the nutrient requirements of a crop has been estimated
quantitatively by Barber et al. (1963) and by Barber and
Olson (1968). Barley (1970) has recently reviewed the subject in considerable detail. When the concentration of a
given nutrient in the soil solution is low, and the demand
by the crop is great, the extent of rooting and root hair
development becomes particularly significant.
    It is well recognized that soil structure, water, aera-
tion, and temperature are major factors in root establishment and growth. The reviews of Wiersma (1959), Pearson
(1966), and Danielson (1967) are adequate in this regard
and further elaboration at this point seems unwarranted.

It must be emphasized, however, that through its effect on root development of the crop, the soil physical condition also plays a most important role in determining nutrient availability and uptake. Proper nutrition requires a continued absorption of nutrients to meet the demands of the rapidly growing crop. This rate process is the limiting factor when nutrition is inadequate, but the specific flux required at any point in the soil is decreased in magnitude when the root surface area capable of uptake is increased.

Rooting depth, as well as degree of branching, is determined both by the genetic characteristics of the species and by the environmental conditions influencing the entire crop plant. In the majority of cases the greatest contribution to the nutrient supply and plant accumulation is made by the surface soil. When the soil fertility level is critically low, however, the nutrient supply from greater depths may be of considerable consequence to the crop. Subsoil structure, water, and aeration then become significant. Subsoil tillage for deeper root penetration has resulted in beneficial, but frequently only temporary, results. We can expect the importance of root system configuration and volume of soil exploited to be greatest for the least mobile nutrients such as phosphorus.

B. Movement to Root Surfaces

The second factor of importance in maintaining adequate nutrient concentration at the root surface is the rate of movement of the nutrients through the soil. This aspect of nutrient availability has long been recognized but has only recently been considered in a quantitative manner. The transport process involves both convection of salts present in solution and ionic diffusion in solution or along charged particle surfaces.

Convective or mass-flow transport occurs as soil water flows due to potential gradients resulting from surface evaporation, irrigation, or transpiration by the crop. The magnitude of the convective flux of nutrients is related to the volume flux of water flow and to the concentration of the nutrients. Diffusive flow results from concentration gradients in the soil solution set up by root absorption or rejection, or by the release of nutrients resulting from organic matter decomposition or fertilizer application.

In recent years, a number of authors have given

consideration to solute movement in relation to crop nutrition. Important among these are Fried and Shapiro (1961), Barber (1962), Barber et al. (1963), Gardner (1965), and Olsen and Kemper (1968).

Evaluation of nutrient movement in the vicinity of plant roots is difficult. Much of the verification of concentration gradients has resulted from autoradiographs showing depletion or accumulation of radioactive nutrient isotopes around individual roots (Barber, 1962). Calculations of the magnitude of transfer by convection have most commonly involved knowledge of the transpiration flux and estimates of the average soil solution concentration. As an example, the computations of Barber et al. (1963) indicate that 500% of the required calcium, 10% of the required potassium, and 1.0% of the required phosphorus could be supplied to corn root surfaces by convection. Average values were used for measured plant and soil solution concentrations and it was assumed that 500 units of water are transpired per unit of dry matter produced. Caution must be used concerning the use of such simplified computations because of the hydrodynamic dispersion resulting when diffusion and convection occur simultaneously and because nutrient uptake and water uptake are not always consistently related over time and space. Although measurements of solute diffusion in soil have not been numerous, the available values of diffusion coefficients in soil water systems are in most instances a few orders of magnitude lower than those for dilute aqueous solutions in bulk.

Olsen and Kemper (1968) have provided an extensive theoretical discussion of the various factors involved in ion diffusion and transport due to convection in a soil system. They have further provided solutions to the governing partial differential equations when certain boundary conditions are assumed. It is impossible to set up mathematical boundary conditions that portray field conditions of the plant-soil-solution-ion system with complete accuracy; a better understanding of the importance of specific concepts and processes can be obtained through such analyses.

At the present time we are concerned with ion transport processes only in connection to how they are influenced by the physical condition of the soil. Using the notation of Olsen and Kemper (1968), the steady state diffusion of nutrient ions may be expressed as

$$\frac{\partial Q}{\partial t} = -[D\theta(\frac{L}{L_e})^2 \ \alpha\gamma]A \ \frac{\partial c}{\partial x} \tag{2}$$

where $\partial Q/\partial t$ is the diffusive flux, D is the diffusion coefficient for the ion in dilute aqueous solution, A the cross-sectional area normal to the direction of net ion movement, and θ the water content of the soil expressed as a decimal fraction. $(L/L_e)^2$ represents a tortuosity factor where L_e is the actual distance of diffusion between two points separated by the straight line distance L. The α term is a viscosity factor caused by adsorption of water to solid surfaces, and γ is a term to account for the influences of pore-size variability along the diffusion path because of the different effects of negative adsorption of anions in pores of various sizes. Finally, $\partial c/\partial x$ is the concentration gradient in the soil solution along the direction of net movement.

The terms grouped in brackets in equation (2) may together be considered the diffusion coefficient, D_p, for the ion in a given soil at a specific water content as long as A is the cross-sectional area of the bulk soil and c is the concentration of the ion in the soil solution. The influence of soil physical conditions, particularly water content and pore configuration, are generally exerted mainly on the factors $\theta(L/L_e)^2 \ \alpha\gamma$.

C. Relative Importance of Root Extension and Nutrient Movement

The relative importance of root extension, convective flow, and diffusion in maintaining the concentration of a specific nutrient at the absorption surface is difficult to assess. Even for a given condition of plant and soil, it is certain to vary diurnally and spacially throughout the profile of the rooting zone. Barber and Olson (1968) have attempted to calculate the contributions of these three processes using some rather simple but perhaps the best available assumptions. A few examples of their calculations are given in Table 1. They have assumed that corn roots occupy 1% of the volume of the surface soil and therefore intercept about 1% of the available nutrients. Mass flow was calculated from average nutrient concentrations of soils in their region and an assumed water use by the crop of 3 million pounds of water per acre.

201

Diffusional contribution was obtained by the difference between requirement and the sum of the root interception and
mass flow contributions. Such generalizations help in
showing the great differences to be expected between nutrients but must be recognized to be, at best, merely approximations.

TABLE 1

The Relative Significance of Root Interception, Mass-Flow,
and Diffusion in Supplying Corn with its Nutrient
Requirements from a Typical Fertile Silt Loam Soil
(Barber and Olson, 1968)

Nutrient	Lbs per acre needed	Approximate amount supplied by		
		Root interception	Mass-flow	Diffusion
N	170	2	168	0
P	35	1	2	32
K	175	4	35	136
Ca	35	60	150	0
S	20	1	19	0
Mo	0.01	0.001	0.02	0

Nutrient Uptake and Use by Plants

The transfer of nutrient elements from the soil solution into the root, once they are in the proper chemical
form and oxidation state and are at the root surface, is
governed both by the physics of the system and by the physiological activity of the plant. This transfer must follow
the general transport law given by equation (3) where the
flux, J, is proportional to a driving force, I, and the
proportionality factor, K, is related to the permeability
of the conducting medium:

$$J = KI \qquad (3)$$

The driving force is a complex factor which may involve
pressure or concentration gradients as well as the energy
released in cell metabolism. Much has been written on the
theory and experimental evidence regarding the relative

importance of the so-called "passive" and "active" modes of solute entry into plants. Steward and Sutcliffe (1959), Fried and Shapiro (1961), and Hendricks (1966a) are excellent examples. The permeability of the root tissue to nutrient entry and translocation varies with time and location in the plant and is to a great extent determined by the properties of cellular membranes.

The complexity of the root permeability problem has been reviewed in considerable detail by Collander (1959). Whether nutrient uptake is governed by physical or biological phenomena, it can be influenced by both internal and external factors. These factors can, in turn, be influenced by the physical conditions of the soil. The actual mechanisms of nutrient entry are only important to this discussion in so far as they affect quantitative relation of uptake rate to solution concentration. Regardless of the active or passive nature of the uptake process, the nature of the cellular constituents influences the uptake through the permeability factor, K, in equation (3). In the case of truly active uptake resulting from metabolic energy release, it is difficult to establish to what extent environmental factors influence permeability per se or on the driving force generated by the metabolism. Hendricks (1966b) discussed the evidence of barrier location in the salt absorption process. He expressed an opinion that the rate-limiting region is the external cytoplasmic membrane, plasmalemma, of the epidermal cells of the root. Collander (1959) implied that the driving force in active transport should be more susceptible than the permeability to external and internal influences.

A number of investigations have shown that the apparent permeability to salt varies with age of the tissue, and in general is higher near the root tip than elsewhere along the root. The work of Kramer and Wiebe (1952) indicates considerable variation in nutrient accumulation and possibly uptake rate along each root and between roots for plants growing under identical conditions. Relatively heavy uptake apparently occurs in the root hair zone one or more centimeters behind the root tip.

The influence of soil physical conditions on cell permeability is difficult to assess. The reports of uptake measurements where temperature and oxygen pressure were varied show significant effects, but the direct cause remains largely hidden. According to Collander (1959),

temperature influences tissue permeability as well as cell
activity. Some effect is likely to be due to changes in
physical properties of water and in coefficients of diffusion. The temperature coefficient values (Q_{10}) measured
for ion absorption and accumulation, however, are high, and
this indicates the significance of the physiological effect.
In contrast, Wallace (1963) reported on uptake from nutrient solution by intact bush beans at 5°C and 25°C. The
Q_{10} values were only slightly greater than unity at high
ion concentrations of 10^{-1} molal while the temperature de-
pendency was much greater at a low concentration of 10^{-4}
molal. His data also indicate an appreciable difference
between ion types.

The dependence of nutrient absorption by plants on soil
aeration, or root oxygen supply, is well known (Grable,
1966). Ion accumulation is linked to aerobic metabolism.
Substrates within the mitochondria are oxidized by molecular oxygen, and the energy so released is transformed in
part to chemical bond energy ATP where it may be used in
nutrient uptake (Hendricks, 1966b). Thus, soil aeration is
important in determining the driving force term of equation
(3). The permeability, according to Collander (1959), does
not appear to be greatly affected by anaerobic conditions.

Soil water potential may exert an influence on both
passive and active absorption. Under sufficiently dry conditions the transpirational flux is decreased and, if solution concentration is limiting, fewer nutrient atoms may be
swept into the root system. Under such conditions plant
growth is restricted, however, and the requirement for nutrients is undoubtedly diminished. At low soil water potentials the root water potential can also be expected to
decrease. The effects of moisture stress on membrane permeability and metabolic activity are not well known.
Brouwer (1954) performed some interesting experiments relating xylem suction to ion uptake and concluded that high
xylem suction caused an increased root conductivity for anion absorption. The conditions of his plants were quite
artificial, however.

Managing the Soil Physical Condition for Improved Plant Nutrition

Considerable information and new methods of measurement
are yet needed before management of the soil physical

condition, aimed at improving the chemical environment of the root system, approaches an exact science. Most of the farming practices involving soil manipulation are carried out for specific purposes such as loosening the soil for subsequent operations, water management, erosion or weed control, incorporation of residues, etc. These practices undoubtedly affect nutrient availability and uptake by plants, but the manipulation of a soil in a specific manner for the express purpose of promoting plant nutrition is rarely observed. Precision measurement of the micro-environment at root surfaces is still largely impossible despite recent advances in the development of micro oxygen electrodes, micro thermistor thermometers, needle penetrometers, and micro-micro techniques for measuring solution or gas compositions. Even when environmental conditions at point sites become measurable, interpretation of the influence of cultural practices will be difficult because of complicated interactions and continuous changes with time.

A. Management of Soil Moisture

The content of water in the rooted soil profile varies throughout the season, gradually decreasing in quantity as evapotranspiration proceeds and then abruptly increasing as precipitation or irrigation takes place. Description of the water regime experienced by the crop requires knowledge of how soil water content and its potential vary with time and with space. In nutritional considerations both water content and matric potential are important. In general, it is to be expected that the greatest utilization of soil and fertilizer nutrients will be achieved if the soil water content is maintained as high as possible without causing aeration or temperature problems.

Adequate soil moisture allows for potential transpiration by the crop. Nutrients are swept to the root surface by the convective flow and are also, to some degree, swept into the root and up the xylem vessels. Lopushinsky (1964) demonstrated appreciable increases in the phosphorus and calcium contents of tomato plants due to forcing water through the plant by increasing the hydrostatic pressure of the solution surrounding the roots. High soil water content also enhances the diffusional transport process, whereas soil drying results in a decrease of the effective cross-sectional area of water through which diffusion can occur. In addition, the effective distance of diffusion, L_e, increases and the significance of the viscosity effects, α, and the negative adsorption effects, γ, are magnified (see equation (2)). The work of Porter et al. (1960) indicates that the product $(L/L_e)^2\alpha\gamma$ is linearly related to water content θ. Thus, as the content of water in the soil decreases, the value of D_p would be expected to decrease in a hyperbolic manner.

Diffusion coefficients for some ions have been measured in relation to the soil water content. A tabulation of pertinent values is given in Table 2 where, in some cases, values were estimated from published curves. In order to emphasize the differences between nutrient types, some of the data in Table 2 are presented in graphical form in Figure 2. The influence of water content is greater for ions not adsorbed to the colloidal surfaces. It must be noted that the D_p values used in Table 2 and Figure 2 are defined by the term in brackets in equation (2). Care must be exercised when comparing effective diffusion coefficients reported in the literature because their definitions are not consistent. The values summarized by Gardner (1965) are equivalent to those of Table 2 divided by θ. It seems more reasonable to consider the influence of soil water content in terms of D_p as presently defined.

The change in transmission properties as soil dries should be related to the actual supply of a nutrient reaching the surface of a root by diffusion. It must be remembered, however, that as θ changes, a change in concentration of the nutrient in solution may result. For anions of completely soluble salts, the concentration in solution is inversely proportional to the water content. In such cases the decrease in magnitude of D_p resulting from drying may be partially offset by increased values for the concentration gradient, $\partial c/\partial x$. This would be expected for chloride and nitrate and would tend to reduce the effect of increased diffusion coefficients with increasing water content as shown in FIgure 2.

Another important consideration, often neglected in theoretical considerations of nutrient diffusion to roots, is that the value of θ decreases steeply toward the root surface when transpiration is taking place. The values for D_p measured in soil cells in the laboratory at an average soil water content may be considerably greater than that near the root of a transpiring plant growing in the same soil at the same average water content.

TABLE 2
Measured Diffusion Coefficients in Soils at Various Levels of Soil Water

Soil	Nutrient	Matric suction bars	Water content % by volume	$D_p \times 10^7$ cm^2 sec^{-1}	Reference
Loam	Cl	0.33	23	8.9	Porter et al.
		1.00	19	5.6	(1960)
		15.6	11	0.6	
Clay	Cl	0.33	43	24.7	
		1.00	35	13.6	
		15.00	24	4.4	
Sandy	Rb		41	0.8	Graham-Bryce
			33	0.2	(1963)
Loam	Rb	0.16	47	4.4	Patil et al.
		1.17	28	1.4	(1963)
		15.00	17	0.6	
Silt (separate)	NO_3	<0.05	43	56.8	Romkens and
		~0.30	36	27.4	Bruce (1964)
		~0.33	28	5.6	
Silty clay loam	P	0.10	42	6.2	Olsen et al.
		0.33	35	3.2	(1965)
		1.00	29	1.9	
		6.00	19	0.4	
Clay	P	0.33	54	5.4	
		1.00	46	3.2	

Evaluations of the influence of specific factors on nutrient uptake by plants usually have involved measurements of nutrient accumulation in the plant tissue, often only in the aerial portion, or of the concentration of the nutrient in the tissue. The 24-hour absorption of Rb by the roots of young, non-transpiring corn seedlings (Figure 3) has been shown by Danielson and Russell (1957) to decrease rapidly as the matric suction of the soil increased through the range common to field conditions. The concentration of

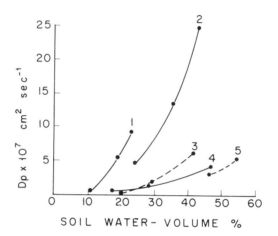

Figure 2. The influence of soil water content on measured self-diffusion coefficients of nutrients in soils.

1. Cl, loam soil   Porter et al. (1960)
2. Cl, clay soil   Porter et al. (1960)
3. P , silty clay loam   Olsen et al. (1965)
4. Rb, loam soil   Patil et al. (1963)
5. P , clay soil   Olsen et al. (1965)

Rb^{86} in the seedlings was in an almost perfect straight line relation to the soil water content. An important fact is that when uptake was measured from inert media at constant water content, but varying in solute potential as a result of mannitol in solution, the Rb uptake from solution was not influenced by solute suction even though root tissue hydration varied markedly with total potential as it did in the studies using soil. It appears, then, that variation in diffusion was the important factor in uptake.

Olsen et al. (1961), using the same technique with four other soils, found that relative phosphorus absorption was related to matric suction in a manner almost identical for each soil and very similar to that of relative Rb absorption (Figure 3). Absorption of Rb by corn roots over a 48-hour period was studied by Place and Barber (1964). In order to minimize the root growth variable they placed the Rb^{86} in a band in the soil. Their results, not unlike those in Figure 3, showed a two-fold increase in uptake as soil moisture status changed from the 15-bar percentage to the moisture equivalent.

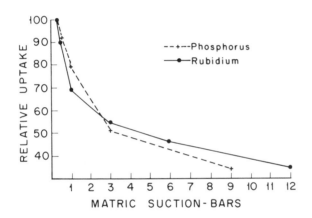

Figure 3. Relative uptake of nutrients by corn seedlings in relation to soil water matric suction [Danielson and Russell (1957) and Olsen et al. (1961)].

Field confirmation of the importance of soil moisture in crop nutrition is difficult to obtain and interpret. Various studies have shown increased uptake of particular nutrients due to added water; however, confounding between nutrient yield and dry matter yield is always a possibility. A typical example is the work of Power et al. (1961) in which the uptake of phosphorus by wheat, under semi-arid conditions of moisture limitations, was increased by additional water. The uptake of P was a linear function of the amount of water used by the crop, which for each treatment was essentially all the water available to it. Thus, interpretation is complex.

Viets (1967) reviewed nutrient availability in relation to soil water and pointed out that the measured influence of water content on diffusion coefficients or on rate of nutrient uptake at specified soil water conditions may not help in predicting the adequacy of plant nutrition resulting under a seasonal water regime. Uptake by the crop during periods of high soil water potential may completely obscure slower nutrient uptake during dry spells. Roots apparently have the capacity to absorb nutrients in excess of their needs. If the plant can supply its needs during wet periods, then decreased availability during dry periods may be of little consequence. However, the real importance of managing the soil for improved nutrition is when limiting conditions exist, rather than when fertility is high.

Of special importance in soil water management for crop nutrition is the maintenance of an adequate supply in the part of the profile where nutrients are most abundant and hence transpirational and diffusional flow to root surfaces can be most effective. Maximum fertility normally resides in the surface zone of the soil but in some instances may occur at greater depths. Crop nutrition may be jeopardized by a dry surface soil even though sufficient water is available deeper in the profile to maintain the transpirational requirement of the crop. This is perhaps a reason why sprinkler irrigation appears in some studies to be superior to furrow irrigation, in which nitrogen may migrate to the surfaces of the ridges where rapid drying takes place. Hodgson (1970) reported increased benefits to grass when fertilizers were placed in the subsoil instead of in the surface soil where drying occurred more frequently.

Water management under conditions of submergence, such as used in rice production, can also affect nutrients. Fluctuations in water content of the soil as a result of flooding and draining the field may create ideal conditions for nitrogen losses by denitrification (Patrick and Mahapatra, 1968). Mineralization of organic matter and subsequent nitrification to nitrate can occur in the oxidized soil surface layer during drained periods. Leaching to lower anaerobic zones, or reflooding of the soil, results in denitrification and loss of the nutrient. Continued submergence prevents nitrate formation and keeps the nitrogen in the ammoniacal form even where denitrification is prevented. At the same time, both native and applied phosphorus is maintained in a more available condition by the unique conditions of the waterlogged soil.

The importance of water control to prevent nutrient removal from the soil root zone by erosion or excessive leaching is well known and requires no further elaboration. Temporary removal of nutrients, especially nitrogen, to regions below the roots of young seedlings is a common cause for deficiency symptoms early in the crop season. Prevention is difficult except when the problem results from the application of too much irrigation water.

B. Management of Soil Aeration

The importance of soil aeration in crop nutrition under normal upland culture is largely related to microorganism activity and root respiration. The supply of air in the soil varies inversely with water content. Herein lies the dilemma: high water content is beneficial to nutrition in many ways as seen in the previous section, but high air content is conducive to soil aeration. Hence, a compromise is required between water and air supply to the root zone. Aeration restrictions in the field are often temporary in nature and the problem may go undetected in many cases.

Evaluation of soil aeration in terms of crop response has not been worked out in a completely satisfactory manner. Letey and his coworkers, at the University of California, Riverside, have made great progress in development of microelectrodes for measuring oxygen diffusion rates in soils (e.g. Stolzy and Letey, 1964), and have established critical minimum values for crop growth. The technique requires a relatively large number of determinations and there has been some doubt as to the meaning of the values obtained under certain conditions (McIntyre, 1970; Dasberg and Bakker, 1970).

The coefficient of gaseous diffusion, similar to that for nutrient diffusion, is the proportionality factor relating the diffusive flux to the concentration gradient. Gaseous diffusion coefficients, however, are not so dependent upon the pore space configuration. Under unsaturated conditions, soil water is contained in the smaller pores, leaving the larger ones for gaseous diffusion. The ratio of the coefficient of gas diffusion in soil to that in bulk

air is closely approximated, over the air porosity range
common to field soils, by a factor of 0.6 times the volume
fraction of air-filled pores. Although the diffusion of
oxygen and carbon dioxide through soil pores is of utmost
importance in microbial and root activity, measurements of
their coefficients have not been of real significance in
evaluating soil aeration. The diffusion coefficient of
oxygen in water is 10^4 times lower than in air. The limiting factor in soil aeration, then, may be the rate of oxygen movement through water films. When the solubility of
oxygen in water is taken into consideration, the transport
rate through the soil solution is of the order of 10^5 times
slower than through air. These factors support the use of
the platinum wire microelectrode in aeration evaluation.

Studies of soil aeration effects have to a large extent
been concerned with nutrient uptake in relation to aeration
levels imposed under controlled conditions. In the field,
any restriction of aeration is likely to be temporary and
difficult to diagnose, and its relation to nutrition rather
complex. Anaerobic conditions may promote the availability
of Fe, Cu, Mo, Mn, and other nutrients, thus partially off-
setting their detrimental effect on root respiration.
Hodgson (1963) has reviewed the complex behavior of micro-
nutrients in soil. Variations in oxidation-reduction potential and pH resulting from changes in aeration due to
water fluctuations add to the complexity.

Management of soil aeration is limited almost entirely
to control of soil water and maintenance of adequate soil
porosity. The limiting value of oxygen in the soil pore
space for rubidium uptake by corn seedlings is higher when
the soil is wetter (Danielson and Russell, 1957). This is
probably due both to the greater need for nutrients because
of increased growth rate, and to the impeding effect of the
thicker moisture films on diffusion to root surfaces. Mac-
ropore volume in the soil is therefore important if low
soil water suction values are to be maintained without
aeration problems. Currie (1962) studied macrostructure
and microstructure aeration and emphasized the importance
of optimizing the size of soil crumbs to promote aeration.
He admitted the difficulty of evaluating the optimum size,
however, and pointed out that it probably changes considerably through the growing season.

Adequate drainage, prevention of soil compaction, and
surface crust control are probably the most practical means

of soil aeration management for improved plant nutrition.

C. Management of Soil Temperature

Temperature can be readily and accurately measured, and control of temperature in experiments designed to evaluate its influence on the soil-plant system is relatively easy to accomplish. In the field, however, it is not so easy to control the temperature accurately. Relatively few commonly used management practices influence soil temperature very much, yet it is probable that the nutrition and resultant yields of commercial crops could be more nearly maximized by controlling soil temperature than by controlling any other single soil physical factor.

The literature is replete with reports of temperature effects on crop response. Nevertheless, an accurate program to control soil profile temperature throughout the growing season cannot be written because no single temperature is optimal from the chemical, biological and physical standpoints. Nielsen and Humphries (1966) have reviewed the work published since the earlier review of Richards et al. (1952).

Nutrient availability and uptake are subject to soil temperature effects through all phases of equation (1). Microbial activity, solubility, ionic equilibrium with soil adsorption sites, diffusion coefficients, and root absorption factors of tissue permeability or metabolic energy supply, are all influenced by soil and root temperature. The temperature dependence of the biological activities in the soil (roots and microorganisms) is undoubtedly much greater than that of the physical processes. The measured rubidium diffusion coefficient in soil, however, was shown by Graham-Bryce (1963) to be appreciably temperature dependent (Table 3). Studies in which the measurement of nutrient accumulation in plants has been related to soil temperature are so numerous and varied that overall interpretation is difficult. In general the biological activities in the soil are expected to increase with temperature to maximum levels in the vicinity of 30°C. Actual absorption of nutrients by plants appears to have temperature dependence curves varying widely for the different nutrients.

The carefully controlled work of Walker (1969) has shown that the rate of dry weight accumulation by maize seedlings reaches a sharp maximum at 26°C soil temperature

TABLE 3

Self Diffusion Coefficients for Rb in Soil
as Influenced by Temperature and Soil Compaction
(Graham-Bryce, 1963).

Bulk density (g cm^{-3})	Temperature (°C)	Diffusion coef. $(\text{cm}^2 \text{ sec}^{-1} \times 10^8)$
1.34	3.5	0.28
1.34	21.0	1.8
1.64	3.5	5.8
1.64	21.0	8.1

(with foliar temperature maintained at 25°C) and decreases rapidly as the soil is cooled or heated from that temperature. Nutrient uptake, expressed as concentration in the shoots, had various response curves. Nitrogen concentration increased with soil temperature from 12° to 18°C, decreased from 18° to 26°C, and increased again from 26° to 34°C. The phosphorus concentration decreased as soil temperature increased from 12° to 25°C and then increased again from 25° to 34°C. Potassium increased in concentration by over 100% as soil temperature was raised from 12° to 18°C and then remained about the same at higher temperatures. Many other elements indicated minimum concentration values at soil temperatures between 20° to 27°C. The need for care in presenting and interpreting data involving other factors is thus emphasized by the appreciable temperature dependence of nutrient absorption, as observed by Walker. The nutrient absorption response to temperature is complex, however, and not all results are consistent. As an example, Knoll et al. (1964) reported phosphorus uptake and accumulation by corn to increase as soil temperature increased from 15° to 25°C. Power et al. (1970) attribute some of the inconsistencies to their finding that temperature influences vary with stages of crop development.

The need to manage soil temperature in the field is too often neglected. Surface soil temperature in early spring can be increased by ridging, orientation of crop rows, or use of mulch. In areas of excessive temperatures, organic mulching can be used to lower soil temperature. References

to these practices may be found elsewhere (Danielson, 1967).
Irrigation is also a means of soil temperature control.
The use of water for cooling or heating is practiced and
the latter may become more significant as power generation
by nuclear reactors becomes more common.

D. Management of Soil Structure

    An important key to promotion of crop nutrition and
growth lies in the management of soil structure. Soil
structure has a considerable influence on the water, air,
and temperature relations of the soil-plant system. The
management of soil structure, however, is even more complex
than its definition or measurement. Bolt et al. (1959)
have clearly pointed out the problems associated with characterizing soil structure as it relates meaningfully to
plant growth. Similar problems exist, of course, with respect to plant nutrition.
    Direct effects of soil structure on crop nutrition
probably exist mainly in the area of root extension and
soil-root contact. Increased bulk density appears to be
beneficial as far as nutrient absorption is concerned, providing aeration and physical impedance to root growth are
not limiting. Cation diffusion coefficients, measured by
Phillips and Brown (1965), increased in value as bulk density increased from 1.05 to 1.30 g/cm^3 and then decreased
again with greater compaction. Graham-Bryce (1963), found
that in a soil of low clay content, compaction of the soil
increased rubidium D_p uptake, as shown in Table 3. The
gravimetric water content of the soil remained constant so
the volumetric content increased with bulk density but only
by approximately 18%. The effect of compaction on anion
diffusion may not be significant. Field compaction of
loose soil by Passioura and Leeper (1963) greatly increased
benefits from residual manganese applied in previous years.
The effect was apparently due to improved root contact with
soil particles and the soil solution.
    More commonly, however, soil compaction results in re-
duced crop nutrition, and tillage practices for loosening
the soil and prevention of compaction are generally required for proper structure management. Pore size hetero-
geneity is desirable to minimize nutrient leaching when
soils are wetted rapidly. Downward flow through large
pores and lateral movement into the small pores of the

aggregates help to keep soluble nutrients in the surface
soil. Management practices for structure development generally involve tillage, use of perennial fibrous-rooted
crops, and additions of organic matter. Commercial soil
conditioning chemicals are effective in aggregate stabilization but economically prohibitive for field-scale use.
These soil conditioners have, however, allowed evaluation
of aggregation on plant nutrition and growth. In general
the results have been inconsistent and point out the fact
that we do not really know what is an optimum soil structure.

"Tillage for Greater Crop Production" was the title of
a conference held in Chicago, the proceedings of which were
published by the American Society of Agricultural Engineers
(1967). The papers included deal with tillage in relation
to soil environment, nutrient relations, and water management, and thus serve as an important source of references
on the subject.

Conclusion

The physical condition of the soil can greatly influence nutrient availability and uptake by crops. Basic relationships between measurable soil properties and crop nutrition are now understood and can be managed. However,
the direct measurement of the soil physical condition,
evaluation of the measurement in terms of plant nutrition,
and devising specific management practices to optimize
these relationships are items still requiring improved understanding.

Soil is not needed for adequate plant nutrition and
growth. Solution culture is perhaps the most ideal method
for supplying the precise needs of the plant. Agriculture
is, however, based almost entirely on soil culture and in
many lands the physical condition of the soil is deteriorating rather than improving. The agriculturist must attempt
to evaluate his soil, crop, and climate in order to manage
his fields so as to improve that factor expected to be most
limiting in production. In arid or semiarid regions without irrigation, soil management for water conservation is
of primary concern. Tillage designed for leaving crop residues on the soil surface may be highly beneficial for
moisture conservation and wind erosion control, but may not
result in the optimum condition for nutrient release to the

soil solution.

Soil organic matter is an important key to continued nutrition of a crop. Problems arising from the soil physical condition are, in the opinion of the author, often related to organic matter management. Limitations of measurement usually prevent positive correlation of such factors, however.

It is to be recognized that the ideal medium from the standpoint of nutrition may not necessarily provide the other requirements of the crop. The soil must be managed so as to obtain a compromise among the various requirements that is consistent with maximizing total yield or economic return. In the majority of cases the soil physical condition that will provide for the most extensive root growth will probably come close to providing the best crop nutrition.

References

1. Amer. Soc. Agr. Eng. (1967). Tillage for greater crop production. Conf. Proc. ASAE, Publ. PROC-168, St. Joseph, Mich.
2. Barber, S.A. (1962). A diffusion and mass-flow concept of soil nutrient availability. Soil Sci. 93, 39-49.
3. Barber, S.A. and Olson, R.A. (1968). Fertilizer use on corn. In "Changing Patterns in Fertilizer Use" (L.B. Nelson et al., eds.). Soil Sci. Soc. Amer., Inc., Madison, Wisc.
4. Barber, S.A., Walker, J.M., and Vasey, E.H. (1963). Mechanisms for the movement of plant nutrients from the soil and fertilizer to the plant root. J. Agr. Food Chem. 11, 204-207.
5. Barley, K.P. (1970). The configuration of the root system in relation to nutrient uptake. Advan. Agron. 22, 159-301.
6. Bolt, G.H., Janse, A.R.P., and Schuffelen, A.C. (1959). Definition and determination of "soil structure." Proc. Int. Symp. Soil Structure, 251-256, Agr. Univ. and Agr. Res. Sta. Bul. XXIV, Ghent, Belgium.
7. Bould, C. and Hewitt, E.J. (1963). Mineral nutrition of plants in soils and in culture media. In "Plant Physiology" (F.C. Steward, ed.), Vol. III, pp. 15-133. Academic Press, New York.
8. Brouwer, R. (1954). The regulating influence of

transpiration and suction tension on the water and salt uptake by the roots of intact Vicia faba plants. Acta Bot. Nederlandica 3, 264-312.

9. Collander, R. (1959). Cell membranes: Their resistance to penetration and their capacity for transport. In "Plant Physiology" (F.C. Steward, ed.), Vol. II, pp. 3-102. Academic Press, New York.

10. Currie, J.A. (1962). The importance of aeration in providing the right conditions for plant growth. J. Sci. Food Agr. 13, 380-385.

11. Danielson, R.E. (1967). Root systems in relation to irrigation. In "Irrigation of Agricultural Lands," (R.M. Hagan et al., eds.), pp. 390-424. Agronomy 11, Amer. Soc. Agron., Madison, Wisc.

12. Danielson, R.E. and Russell, M.B. (1957). Ion absorption by corn roots as influenced by moisture and aeration. Soil Sci. Soc. Amer. Proc. 21, 3-6.

13. Dasberg, S. and Bakker, J.W. (1970). Characterizing soil aeration under changing soil moisture conditions for bean growth. Agron. J. 62, 689-692.

14. Fehrenbacher, J.B. and Snider, H.J. (1954). Corn root penetration in Muscatine, Elliott, and Cisne soils. Soil Sci. 77, 281-291.

15. Fried, M. and Broeshart, H. (1967). "The Soil-Plant System in Relation to Inorganic Nutrition." Academic Press, New York.

16. Fried, M. and Shapiro, R.E. (1961). Soil-plant relationships in ion uptake. Ann. Rev. Plant Physiol. 12, 91-112.

17. Gardner, W.R. (1965). Movement of nitrogen in soil. In "Soil Nitrogen" (W.V. Bartholomew and F.E. Clark, eds.), pp. 550-572. Agronomy 10, Amer. Soc. Agron., Madison, Wisc.

18. Grable, A.R. (1966). Soil aeration and plant growth. Advan. Agron. 18, 57-106.

19. Graham-Bryce, I.J. (1963). Self diffusion of ions in soil: I. Cations. J. Soil Sci. 14, 188-194.

20. Hendricks, S.B. (1966a). Nutrient transfer and plant absorption mechanisms. In "Plant Environment and Efficient Water Use" (W.H. Pierre et al., eds.), pp. 150-176. Amer. Soc. Agron. and Soil Sci. Soc. Amer., Madison, Wisc.

21. Hendricks, S.B. (1966b). Salt entry into plants. Soil Sci. Soc. Amer. Proc. 30, 1-7.

22. Hodgson, D.R. (1970). Some recent experiments on the use of nitrogen fertilisers on grass with special reference to ammonia and the availability of nitrogen in dry conditions. In "Nitrogen Nutrition of the Plant" (E.A. Kirkby, ed.), pp. 201-214. Univ. of Leeds, Leeds, England.

23. Hodgson, J.F. (1963). Chemistry of the micronutrient elements in soils. Advan. Agron. 15, 119-159.

24. Holliday, R. (1970). Soil profile moisture and nitrogen availability. In "Nitrogen Nutrition of the Plant" (E.A. Kirkby, ed.), pp. 189-200. Univ. of Leeds, Leeds, England.

25. Knoll, H.A., Brady, N.C., and Lathwell, D.J. (1964). Effect of soil temperature and phosphorus fertilization on the growth and phosphorus content of corn. Agron. J. 56, 145-147.

26. Kramer, P.J., and Wiebe, H.H. (1952). Longitudinal gradients of P^{32} absorption in roots. Plant Physiol. 27, 661-674.

27. Lopushinsky, W. (1964). Effect of water movement on ion movement into the xylem of tomato roots. Plant Physiol. 39, 494-501.

28. McCalla, T.M. (1959). Microorganisms and their activity with crop residues. Nebr. Exp. Sta. Bul. 453, 1-31.

29. McIntyre, D.S. (1970). The platinum microelectrode method for soil aeration measurement. Advan. Agron. 22, 235-283.

30. Mitchell, R.L. (1970). "Crop Growth and Culture," The Iowa State Univ. Press, Ames, Iowa.

31. Nielsen, K.F. and Humphries, E.C. (1966). Effects of root temperature on plant growth. Soils and Fertilizers 29, 1-7.

32. Olsen, S.R. and Kemper, W.D. (1968). Movement of nutrients to plant roots. Advan. Agron. 20, 91-151.

33. Olsen, S.R., Watanabe, F.S., and Danielson, R.E. (1961). Phosphorus absorption by corn roots as affected by moisture and phosphorus concentration. Soil Sci. Soc. Amer. Proc. 25, 289-294.

34. Olsen, S.R., Kemper, W.D., and Van Schaik, J.C. (1965). Self-diffusion coefficients of phosphorus in soil measured by transient and steady-state methods. Soil Sci. Soc. Amer. Proc. 29, 154-158.

35. Page, J.B. and Bodman, G.B. (1951). The effect of soil physical properties on nutrient availability. In

"Mineral Nutrition of Plants" (E. Truog, ed.), pp. 133-166. Univ. Wisc. Press, Madison, Wisc.

36. Passioura, J.B. and Leeper, G.W. (1963). Soil compaction and manganese deficiency. Nature 200, 29-30.

37. Patil, A.S., King, K.M., and Miller, M.H. (1963). Self-diffusion of rubidium as influenced by soil moisture tension. Can. J. Soil Sci. 43, 44-51.

38. Patrick, W.H. and Mahapatra, I.C. (1968). Transformation and availability to rice of nitrogen and phosphorus in waterlogged soils. Advan. Agron. 20, 323-359.

39. Pearson, R.W. (1966). Soil environment and root development. In "Plant Environment and Efficient Water Use" (W.H. Pierre et al., eds.), pp. 95-126. Amer. Soc. Agron. and Soil Sci. Soc. Amer., Madison, Wisc.

40. Phillips, R.E. and Brown, D.A. (1965). Ion diffusion. III. The effect of soil compaction on self diffusion of rubidium-86 and strontium-89. Soil Sci. Soc. Amer. Proc. 29, 657-661.

41. Place, G.A. and Barber, S. (1964). The effect of soil moisture and rubidium concentration on diffusion and uptake of rubidium-86. Soil Sci. Soc. Amer. Proc. 28, 239-243.

42. Porter, L.K., Kemper, W.D., Jackson, R.D., and Stewart, B.A. (1960). Chloride diffusion in soils as influenced by moisture content. Soil Sci. Soc. Amer. Proc. 24, 460-463.

43. Power, J.F., Reichman, G.A., and Grunes, D.L. (1961). The influence of phosphorus fertilization and moisture on growth and nutrient absorption by spring wheat. II. Soil and fertilizer P uptake in plants. Soil Sci. Soc. Amer. Proc. 25, 210-213.

44. Power, J.F., Grunes, D.L., Reichman, G.A., and Willis, W.O. (1970). Effect of soil temperature on rate of barley development and nutrition. Agron. J. 62, 567-571.

45. Richards, S.J., Hagan, R.M., and McCalla, T.M. (1952). Soil temperature and plant growth. In "Soil Physical Conditions and Plant Growth" (B.T. Shaw, ed.), pp. 303-480. Academic Press, New York.

46. Romkens, M.J.M. and Bruce, R.R. (1964). Nitrate diffusivity in relation to moisture content of non-absorbing porous media. Soil Sci. 98, 322-337.

47. Rosenberg, N.J. (1964). Response of plants to the physical effects of soil compaction. Advan. Agron. 16,

181-196.
48. Shaw, B.T. (ed.)(1952). "Soil Physical Conditions and Plant Growth." Academic Press, New York.
49. Steward, F.C. and Sutcliffe, J.F. (1959). Plants in relation to inorganic salts. In "Plant Physiology" (F.C. Steward, ed.), Vol. II, pp. 253-478. Academic Press, New York.
50. Stolzy, L.H. and Letey, J. (1964). Characterizing soil oxygen conditions with a platinum microelectrode. Advan. Agron. 16, 249-279.
51. Sutcliffe, J.F. (1962). "Mineral Salts Absorption in Plants." Pergamon Press, London.
52. Viets, F.G., Jr. (1967). Nutrient availability in relation to soil water. In "Irrigation of Agricultural Lands" (R.M. Hagan et al., eds.), pp. 458-471. Agronomy 11, Amer. Soc. Agron., Madison, Wisc.
53. Wadleigh, C.H. (1949). Mineral nutrition of plants. Ann. Rev. Biochem. 18, 655-678.
54. Wadleigh, C.H. and Richards, L.A. (1951). Soil moisture and the mineral nutrition of plants. In "Mineral Nutrition of Plants" (E. Truog, ed.), pp. 411-450. Univ. Wisc. Press, Madison, Wisc.
55. Walker, J.M. (1969). One-degree increments in soil temperatures affect maize seedling behavior. Soil Sci. Soc. Amer. Proc. 33, 729-736.
56. Wallace, A. (1963). "Solute Uptake by Intact Plants." (The Author), Los Angeles, Calif.
57. Watanabe, F.S., Olsen, S.R., and Danielson, R.E. (1960). Phosphorus availability as related to soil moisture. Trans 7th Int. Congr. Soil Sci., Madison, Wisc. 3, 450-456.
58. Wiersma, D. (1959). The soil environment and root development. Advan. Agron. 11, 43-51.

ANALYSIS OF GROWTH PARAMETERS AND THEIR FLUCTUATIONS
IN SEARCHING FOR INCREASED YIELDS

D. Zaslavsky
Technion, Israel Institute of Technology

The Medium We Work with

At the outset we must remind ourselves of some well
known but greatly neglected facts about the system involved
in agricultural production.

The system has a large number of phases and components.
Hence, to describe the system one has to specify a large
number of variables. They involve not only soil variations
in space and time but also climatic conditions, plant biological variables, entomological variables and man-made
ones.

In principle, at least, each of the variables is given
values at points in time and space. Thus, the soil moisture content may be formally described in terms of its dis-
tributions over the field, in depth and in time. Thus can
oxygen content and fertilizer distribution, as well as leaf
area, temperature, porosity, infiltration capacity and
slope, be given values in time, space or both.

The variation in time and space may be divided into two
distinct parts:

    a) A more or less smooth change in time and space.
       Such a smooth change may be specified by a series of
       experimental points or by a formula.

    b) Fluctuations in time or space about the smooth line.
The distinction between the two parts of the variation must
be based upon some acceptable procedure. The first part
which is smooth is obtained by some form of averaging over
the fluctuating value. This averaging process must be well
defined in terms of at least two criteria:

    a) The form of averaging: Usually people tend to form
       an arithmetic average. If the averaged property is
       P over the space A, then

$$\overline{P} = \frac{\int WP\,dA}{\int W\,dA} \tag{1}$$

where W is some weight function. However, more generally we are interested in a function F(P) (e.g., the yield as a function of fertilizer level P) and one has to find $\overline{F(P)}$ which is the average yield rather than $F(\overline{P})$ which is the yield at the average fertilizer level $\overline{P}$. A more systematic treatment of this topic was given by Zaslavsky (1967 and 1970).

b) The scale of averaging: We perform the averaging over a subsystem of a given size (in time and space). In determining this size we actually decide which variations are to be smoothed out and excluded from the written description, and which variations are to be recorded and used as averaged values. In the crop field we may wish to set the size of this subsystem at several rows by several plants. We then record the conditions and the yield over such a plot on the average. The values for each row or each plant are not recorded. Similarly we may wish to consider in detail water streams of a certain size and only average slopes between such streams, thus disregarding minor surface perturbations such as small gullies, soil clods, or single grains.

There is good reason for us to review these well known, seemingly trivial facts about the agricultural medium. We wish to pose two fundamental questions arising from the above:

a) The fluctuations about the averaged or smoothed variables may be described by such statistics as the variance, the third moment, or the frequency spectrum (in time or space cycles). The question is how important is the statistic of such fluctuations in affecting yield or, more generally, in contributing to our understanding of the medium.

b) The number of variables is quite large and hard to measure even in the simplest cases. The above discussion raises two more difficulties. It requires that the measurement be of the proper scale of averaging. Most present day methods for measuring soil, rain, and other climatic variables, are really of too small a scale. Moreover, in cases where the statistics of fluctuations is proven to be of

importance, one has to make enough small-scale measurements to define the statistics adequately. The
question is whether there is any practical way to
obtain all these significant measurements. For example, to date there are very few water vs. yield
curves available. Consider for example the large
number of possible combinations of plant varieties,
soils, and climatic regions in Israel. Even if all
other agrotechnical parameters remain unchanged,
would it be possible or practical to define these
combinations and their quantitative interactions?

Possible Effects of Fluctuations on Plant Yields

There is a number of ways by which fluctuations in a
variable may affect the averaged results. For example, one
may ask how fluctuations in moisture and fertilizer distribution may affect plant yield.

Some recent work (Zaslavsky and Buras, 1967; Zaslavsky
and Mokady, 1966) has shown that the effect may be quite
significant. Consider, for example, a yield function at a
point in the field

$$Y(Q_1, Q_2, \ldots, Q_n) \tag{2}$$

where Y is the yield and $Q_1, Q_2, \ldots, Q_n$ are various growth
parameters such as the levels of moisture and various fertilizers. Each of the values Q_i has an average $\overline{Q}_i$ over the
field (or over time or both) and fluctuations q_i. To obtain the average yield $\overline{Y}$ we expand (2) by Taylor's series
around the values of $\overline{Q}_i$. Then we integrate over the field
area A and divide by the area. The result is, neglecting
terms higher than the second,

$$\overline{Y} = Y(\overline{Q}_1, \overline{Q}_2, \ldots, \overline{Q}_n) +$$

$$\frac{1}{2} \frac{\partial^2 Y}{\partial Q_1^2} \sigma_1^2 + \frac{1}{2} \frac{\partial^2 Y}{\partial Q_2^2} \sigma_2^2 + \ldots \frac{1}{2} \frac{\partial^2 Y}{\partial Q_n^2} \sigma_n^2 +$$

$$\frac{1}{2} \sum_1^n \frac{\partial^2 Y}{\partial Q_i \partial Q_j} \rho_{ij} \sigma_i \sigma_j \tag{3}$$

Here the first term is the yield that would have been obtained with perfectly uniform levels Q_i all at the average

225

values $\overline{Q}_i$. The other terms are the first correction to the yield due to fluctuations. For example, consider the case of moisture fluctuations in space and no other fluctuations. Their variance is σ_1^2 and the only correction term is

$$\frac{1}{2} \frac{\partial^2 Y}{\partial Q_1^2} \sigma_1^2 \tag{4}$$

In the lack of detailed data about $Y(Q_1)$ (water vs. yield curves) one can heuristically estimate the second derivative to be of the order of one unit or few units (in some nondimensional scale). Assume the moisture fluctuations to reach as high as ±20%. Then, the variance σ_1^2 is around 0.01. The change in yield due to a nonuniform water distribution is the product of the variance in growth parameter by the second derivative of the yield with respect to that parameter. It will usually not be beyond a few percent. If the moisture response curve is convex then the second derivative of expression (4) is negative and it will be a yield reduction. Not all fluctuations will affect the yield. There will be a difference between onions and citrus trees, since the latter have some averaging capacity themselves. Fluctuations of high frequency or short cycle (shorter than the root space taken by one tree) will not be included in the variance. In short, we have a tool to estimate the potential of increasing the yield by a more uniform distribution of irrigation water. We find that it is negligible, and perhaps may not be easily measurable in field experiments.

To the contrary of the above example, we can show that fluctuations of soil moisture in time may have a very considerable effect on the yield.

One may carry out a series of hypothetical experiments in which the soil moisture in each case is held constant throughout the growing season. Each experiment is carried out with a different moisture level $\overline{Q}$. At some value $\overline{Q}_{max}$, maximum yield will be obtained. At a lower moisture there is water shortage, whereas at higher moisture ($\overline{Q} > \overline{Q}_{max}$) there is an aeration problem due to waterlogging. The analysis of temporal fluctuations in moisture indicates the following. At moisture levels below $\overline{Q}_{max}$ maximal yields are obtained with no moisture fluctuations. In other words, the best policy is to have the most frequent water application or at best a continuous one. On the other hand, in soils with aeration problems or with moisture contents higher than Q_{max} it is better to have some moisture fluctuations with time. In other words, irrigation intervals should be large enough to allow for a certain degree of moisture depletion. Actual experiments (soon to be published) have indicated that a change from a seven day period to a continuous irrigation can more than double the vegetative yield.

We can now see how using averages only and neglecting the variances may lead to substantial errors. In the case of the space distribution of moisture, we may be led to futile efforts to improve uniformity of moisture distribution. In the different case of temporal fluctuations, we may be led to disregard a simple technology that can more than double our crop yields.

The most obvious and well known example of fluctuations and their effect on yield is probably that of thermal fluctuations. It is quite obvious that the average temperature cannot serve as the only measure of the thermal regime. Thermal fluctuations and their effect can be analyzed by the same method described above.

The subject of this symposium is "Optimizing the Soil Physical Environment toward Greater Crop Yields." Clearly, no such optimization can be achieved completely without taking into account the fluctuations of the physical environment factors about their averages.

Multi-Variate Fluctuations

In the above analysis we cited two examples with fluctuations of one variable only (in our case moisture distribution). A similar analysis may be made for fertilizer distribution. Such an analysis indicates an advantage for uniform fertilizer distribution in some cases, and for a band application in other cases, depending on the sign of the second derivative in term (4).

Equation (3) suggests one form by which experiments and analysis of multivariate fluctuations may be handled. A typical multivariate correction term is

$$\frac{\partial^2 Y}{\partial Q_1 \partial Q_2} \, \rho_{12} \sigma_1 \sigma_2 \tag{5}$$

Q_1 and Q_2 may be the levels of a given fertilizer and the level of moisture, or the levels of two fertilizers. The mixed partial derivative of the yield Y with respect to these levels Q_1 and Q_2 is the mixed response or the inter- action. It may be either positive or negative. σ_1 and σ_2 are the standard deviations of the fluctuations of Q_1 and Q_2, respectively. They are always positive. ρ_{12} is the correlation between the fluctuations of the two. For exam- ple, if the fertilizer is brought in by water, then $\rho_{12} = 1$. If two fertilizers are independently distributed, ρ_{12} may vanish. In case of leaching the correlation between water and fertilizer levels may be negative.

As the interacting response to fertilizers (and hence also the mixed derivative) is known to be most often posi- tive, our analysis clearly calls for coplacement of ferti- lizers. By coplacement ρ_{12} becomes positive and since the whole expression in (5) is then positive it adds to the yield. To get a maximum addition to the yield we wish to have large fluctuations in the fertilizers. In other words, for a given quantity of the fertilizers, we should coplace them in sites of high concentrations, with no fer- tilizers in other parts of the soil. One has of course to be careful not to lose yield by getting another term in equation (3) more negative. Two fertilizers may have a nondimensional mixed derivative of several tens percent. The correlation of ρ_{12} in coplacement may be unity and $\sigma_1\sigma_2$ may be equal to several units. Thus one may expect more than doubling the yield by coplacement of fertilizers in high concentrations. In a similar fashion a negative in- teraction may cause a large reduction in yield.

We have thus indicated a scheme to search for higher yields by a more complete control of environmental varia- bles. This has been done without entering into the intri- cate physical and physiological mechanisms involved. While the study of such mechanisms is both important and inter- esting, it is not a substitute for our algebraic analysis, which provides guidelines in searching for an agricultural optimum. The lack of such simple guidelines has obscured for very long important indications of possible yield in- creases. Yields in hydroponics have often been doubled or trebled as compared with conventional methods of produc- tion. The two obvious properties in hydroponics are high positive correlations between all chemical ingredients and frequent irrigations. It is interesting that the reason given for drip irrigation in Israel was initially the saving of water and not the increase of yields. This is rather ironic in view of the fact that a very simple analysis of unsaturated flow can indicate that a continuous drip irrigation is almost certain to waste water by downward seepage. However, despite a highly uneven moisture distribution in space, yields are increased under drip irrigation. This is quite in line with the above analysis, which has been available here for several years. Disregarding this analysis, the search for better yields has continued to be intuitive, based on long cycles of trial and error and involving major expenditures on intermediate and sometimes useless technologies.

Net Change in Equilibrium Conditions Due to Fluctuations

If the barometric pressure or the temperature fluctuates with time (or space), a given function of them also fluctuates. More specifically, the energy of the water or of any other soil constituent fluctuates. A case in point is that the free energy is usually a nonlinear function of the temperature or pressure. Thus we may consider two points in the soil. Both may have the same average temperature or pressure. However, if one point has a different variance of the temperature and pressure it will have also a different average free energy level, as can be easily shown by the process of averaging. This effect can produce unexpected fluxes in the soil. For example, the daily temperature fluctuates in the soil and the amplitude decays with depth. Thus the variance decreases with depth. As a result, the water in the top soil is at a higher average free energy (because the free energy increases in proportion to the temperature raised to a power higher than unity). There will therefore be a net flux of water although the average temperature is the same throughout the soil profile.

A similar phenomenon may be caused by barometric fluctuations, since the effective volume of the moisture may vary with pressure because of entrapped air and soil elasticity effects.

Relaxation of Hysteresis

Temperature and barometric pressure fluctuations cause

229

temporal fluctuations in the water pressure. The possible effect such fluctuations may have on hysteresis of the soil moisture-suction curve has not been studied. It seems that such fluctuations may induce some relaxation of hysteresis toward a moisture suction curve which is intermediate between the primary wetting and drying curves.

Increased Apparent Conductivity

Molecular diffusion is a very slow process. Temporal fluctuations in pressure and temperature induce also fluctuating mass flows. Let us assume that there is no net flow. Still it may be shown that due to dispersion of solutes that accompanies the mass flow there can be an enhanced movement due to concentration gradients. It is quite possible in some cases that there exists a threshold gradient and flow can start only above a given value. In all such cases fluctuations in the gradient may induce an increased net flow.

It is of great interest to study what is the effect of increased net flow on nutrient distribution and biological activity in the soil. It is well known that two regions of the same average temperature may be quite different, depending on the extremes in temperature, humidity (and perhaps also of barometric pressure) experienced at each region.

It is time we reexamine our "carefully" controlled laboratory experiments where constant temperature and pressure are very well maintained. It may be that we thus miss some of the important processes which occur in the field.

Some Hydrological Conjectures (Zaslavsky, 1970)

Rain intensity is a mode of nonuniform distribution in time and space. Each raindrop constitutes an extreme local and temporal fluctuation. The local microslopes are quite irregular. The hydraulic conductivity of the soil is also nonuniform. Thus the phenomena of infiltration and runoff exhibit very significant effects of fluctuations. It is well established experimentally that runoff commences much before the rain surpasses the infiltration capacity of the soil. In fact it can start at very small and low intensity rains.

To demonstrate this phenomenon in its extreme, consider

an agricultural soil with a hydraulic conductivity that averages around 10^{-4} cm/sec. To reach the average infiltration capacity the rain must exceed 100 mm per day. However, quite obviously, a rain of 100 mm during a single day will produce tremendous runoff in a field of such soil even with a slight slope. Clearly, to neglect the effect of fluctuations in the soil and rain parameters would be tantamount to disregarding the most important effect.

The analysis of fluctuation appropriate to this case is quite different than the one brought above. This is basically because the relation between rain, infiltration and runoff is assumed to be linear. It is interesting that averaging here over part of the fluctuations (e.g. only for points where the rain exceeds the infiltration capacity) produces a nonlinear overall relation between the runoff and the rain.

If the rain is P and the runoff is R then it can be shown from heuristic statistical analysis that

for P → 0 $\quad\quad\quad\quad \partial R/\partial P = \dfrac{1}{2}$

for large P $\quad\quad\quad \partial R/\partial P \to 1$

This is quite in line with the experimental formula suggested by the U.S. Soil Conservation Service.

Conclusions

It has been demonstrated that the effect of many environmental parameters cannot properly be specified only on the basis of their averages. The variances and the correlations of their statistics should also be specified. Failing to include such statistics in our measurements and analysis may obscure some very important processes. In searching for optimum yields we must initiate a more systematic treatment of fluctuation analysis. In some cases it is anticipated that very large yield increases may be gained by relatively simple and cheap changes in agrotechniques. The extremely large number of pertinent parameters and the complexity of their measurement has led us to undertake some heuristic approaches to which fluctuation analysis lends itself.

References

1. Zaslavsky, D. (1968). Average entities in kinematics and thermodynamics in porous materials. Soil Sci. 106, 358-362.
2. Zaslavsky, D. (1969). Temporal fluctuations in soil environment and their physical significance. Soil Sci. 108, 326-334.
3. Zaslavsky, D. (1970). Some aspects of watershed hydrology. U.S. Department of Agriculture, ARS 41-157, March 1970, pp. 9-19.
4. Zaslavsky, D. and Buras, N. (1967). Crop yield response to nonuniform application of irrigation water. Trans. Amer. Soc. Agr. Eng. 10, 196-200.
5. Zaslavsky, D. and Mokady, R.S. (1966). Nonuniform distribution of phosphorous fertilizers: An analytical approach. Soil Sci. 104, 1-6.

SUBJECT INDEX

A

Absorption, mineral, 26
Adhesion, 50, 51
Adsorption, 105, 108, 112
   electrostatic, 50
   negative, 206
Advection, 152, 170
Aeration of soil, 35, 38, 44, 45, 48, 61-
   64, 74, 93, 97, 194, 195, 196, 198,
   199, 204, 205, 211-213, 215,
   226, 227
   management of, 211-213
Aggregate stabilization, 45, 52, 216
Aggregates of soil, 48, 54, 216, *see also*
   Soil crumbs
Aggregation, 49
Agricultural physics, 4
Air entrapped, 229
Air porosity, 212
Albedo, 144, *see also* Reflectance,
   Reflection coefficient, Reflectivity
Anaerobic conditions, 63, 210, 212
Anaerobiosis, 62
Analytical approach, limitations of, 24
Angle of incidence, radiation, 178,
   184-186
Anions, exclusion of, 105, 108, 112
Aquifer(s), 20, 21
Arid regions climate, 82, 101
Arid soils, 53
Asphalt barriers, 19, 35, 36, 38-40
Autoirrigators, 92
Averaging, form and scale of, 223-231

B

Barriers to restrict drainage, 170, *see also*
   Asphalt barriers
Beam radiation, 179
Biodegradation, 20
Biological engineering, 23
Biosphere, 4
Bisociation and creativity, 6
Bitumen, 52
Bonding, 53
Bonds, 46, 50
Bowen ratio, 146, 152
Breakthrough curves, 108, 114
Bulk density of soil, 61, 73, 74, 215

C

Capillary fringe, 155, 157
Capillary rise, 36, 37, 80, 125
Cation exchange capacity, 35, 44, 46,
   47, 53, 54
Cations
   divalent, 111
   monovalent, 111
Clay domain(s), 48, 49
Clay soils, 48, 53
Clay-water system, 105
Climatic data, *see* Meteorological data
Coagulation, 51
Coating of soil and canopies,
   174, 175, 181, 182, 186, 189

Cohesion, 51
Colloids, 110
Compaction of soil, 69, 70, 194, 212,
    214, 215
Computers, uses and methods, 25, 26,
    28, 43, 112-115, 117, 121, 138,
    143, 146, 157, 174
Conductivity
    capillary, 36
    electrical, of solutions, 102, 119, 120
    hydraulic, 15, 61, 86-88, 109-112,
        120, 166, 167, 230, 231
    thermal, 24
Consumptive use, 151
Continuity equation, 107
Continuous systems modelling
    language, 28
Convective flow, viscous flow, mass flow,
    104-108, 114, 199-202, 230
Crop ecology, 23, 27
Crop growth, 26
Crop nutrition, 210, see also plant
    nutrition
Crop physiology, 26
Crop response
    to soil salinity, 120
    to soil temperature, 28-30
Crust of soil surface, 52, 212
Cultivation, 134, see also Tillage

D

Darcy's equation, 109
Denitrification, 210, 211
Desert, sandy, 52
De Wit relationship, 164, 165, 169
Diffusion, diffusion coefficient, 38, 62,
    104-106, 113, 114, 146, 197,
    199-202, 204, 206-208,
    210-214, 230
    of water vapor, 146, 157
Dispersion
    hydrodynamic, 106, 108, 114,
        116, 200
    of soil colloids, 110, 111

Dispersion coefficient, 108
Displacement misible, 119
Double-layer theory, 110
Drainage, 11, 14-17, 19, 36-38, 40, 45,
    53, 80-86, 88, 89, 112, 120, 125,
    147, 153, 154, 165, 170, 210, 212

E

Ecological instability, 4
Ecologists, 4, 5, 7
Ecology, 27
Emulsifiers, 52
Emulsion(s), 47, 51, 53
    bitumenous, 46, 50
    hydrophobic, 48, 52
    hydrophilic, 52
Energy balance, 97, 139, 140, 149
Environment, 5, 31, 53, 84, 92,
    95-97, 194, 205
    controlled, 31
    modification of, 50
Environmental management 20-21
Environmental quality, 21
Environmental variables and
    parameters, 228, 231
Erosion, 20, 21, 195, 205, 211, 216
Evaporation, 12, 13, 19, 24, 35, 51, 52,
    80, 81, 83, 88, 89, 107, 112-114,
    135, 154, 157, 163-170; 199
Evaporation pans, 135
Evaporative demand, 139, see also
    Evapotranspiration
Evapotranspiration, 11, 16, 82-86,
    88, 90, 94, 97, 125, 135, 136,
    139, 143, 145, 146, 148, 153,
    163-166, 170, 205
    combination equation, 139, 144
Extinction coefficient, 173, 174, 176
    potential, 94, 139, 153

F

Fallowing (fallow-land, fallow-soil),
    18, 95, 107, 112, 113, 170

Fertility of soil, 35, 103, 137, 199, 210
Fertilization, 40
Fertilizer(s), 36, 133, 137, 195, 199, 223,
    224, 227, 228
  coplacement of, 228
Fick's law, 104
Field capacity, 85, 92, 93, 102, 118, 153
  effective, 153-155
Field environment, 31
Flocculation of clay, 45, 46
Flux meters, 85
Foliage (canopy) cover density, 173,
    176, 189
Furrow irrigation, 147

G

Germination, 26, 52, 58, 63, 66, 104
Greenhouse(s), 97
Ground water, 20
Growth parameters and fluctuations,
    223-232

H

Heat, sensible, 24
Heat balance, 173
Heat capacity, 24
Humid-region soils, 53
Hydraulic gradient(s) 20, 87, 88, 109,
    110, 167
Hydraulic head, 86, 87, 109
Hydraulic properties of soil, 85, 89
Hydrophilic effect, 46, 50, 52, 54
Hydrophobic effect, 46, 51, 54
Hydroponics, 228
Hysteresis, 36, 112, 229-230

I

Impedance mechanical, 53, 215, *see also*
    Soil strength
  to root growth, 215

Infiltrability, 35
Infiltration, 35, 50, 88, 107,
    112-114, 223, 230, 231
Infiltration capacity, 230, 231
Infrared radiation and reflection, 177,
    179, 183, 184, 188, 190
Instantaneous profile method, 86
Interparticle contact (bonding), 53
Ion exchange, 108, 111, 195
Irrigation, 11, 15-17, 20, 36, 38, 40, 45,
    53, 80-82, 85, 87, 90, 93, 103,
    113, 117, 118, 121, 123-125, 128,
    133-161, 163-165, 170, 196, 199,
    205, 210, 211, 215, 216, 226,
    227, 229
  automation of, 94
  continuous, 45
  development of, 11, 12
  efficiency of, 12-14, 90, 133, 148
  future of, 12
  frequency of, 128
  methods, new, 93
  programming, 133-161
    instruments for, 135
  trickle (drip), 94, 228
Irrigation forecasts, 136, 142
Irrigation management service, 134-136,
    138, 141, 158
Isosalinity curves, 122, 127

K

Krilium, 48

L

Leaching of salts, 15, 16, 35, 85, 93,
    111, 119-121, 123, 124, 148,
    195, 210, 211, 215, 228
Leaching efficiency, 119, 120
Leaching requirement, 16, 17, 124, 125
Leaf area index, 146, 173
Light enrichment, 175

Light scattering by foliage, 189
Lysimeter(s), 20, 84, 90, 152, 164, 165

M

Mass flow, *see* Convective flow
Matric potential, 15, *see also* Suction
Meteorological (climatic) data, 139, 140,
    143, 148, 157, 158
Micelle migration, 50
Microbial activity, 213
Microelectrodes, 211, 212
Microorganisms in soil, 196
Model canopies, 174, 186, 187
Modelling approach, 28
Moisture content, optimal, 52
Moisture equivalent, 209
Moisture profile, 15
Moisture regimes, 43
Moisture stress, 94
Mulch(es), 46, 168, 169, 214
Mulching, 23, 26, 35, 39

N

Neutron moisture probes (meter), 39,
    83, 87
Nitrogen fixation, 196
Numerical simulation, 25
Nutrient availability, 194, 199, 210, 216
Nutrient movement to root surfaces,
    199-201
Nutrient requirements, 202
Nutrient supply and uptake by plants,
    193-221
Nutrient transport, 197
Nutrient uptake, active or passive, 203,
    204

O

Organic matter, 45, 196, 199, 210,
    216, 217
    mineralization of, 196, 210

Osmotic adjustment, 102
Osmotic effects, 102, 103, 109
Osmotic efficiency coefficient, 110
Osmotic gradients, 109, 110
Osmotic pressure, 53, 102, 103, 109
Oxidation-reduction potential, 196, 212
Oxygen diffusion, 38, 62, 211

P

Parkinson's law, corollary, 6
Penetration, 53
Penetrometer resistance, 59, 66-69,
    73, 74
Penman formula for evapotranspiration,
    14, 144, 145, 149, 152
Percolation, 35, 36, 40, 80, 84, 85, 90,
    163
Permeability
    of roots, 203
    of soil, 17, 110, 111, 202, 203, 213
Pesticides, 53, 137
Pfeffer's technique, 58
Photographs
    hemispherical, 179, 189
    time-lapse, 59, 67, 70
Photosynthesis, 28, 91, 173-175
Phytotron, 31
Plant environment, 92, 96-97, 101
Plant growth, 40
Plant growth models, 157
Plant nutrition, 193-195
Plant-soil-water relations, 141
Plastic covered soil, 175
Plowing, 57, *see also* Tillage
Pollution of water and soil, 11, 20, 21, 31
    thermal, 31
Polymer(s), 48, 49, 50, 54,
        see also soil conditioners
    emulsions, 50-53
Polymerization, 49
Pore size distribution, 24, 201
Pore water velocity, 105
Porosity of soil, 212, 223
Potential gradient(s), 36, 86, 199

R

Radiation
absorption by soil, 187
balance, 174-179, 181, 185, 188-190
distribution in canopies, 173
interception of, by canopy, 181
photosynthetically active, 183, 188
solar, 24, 144, 145, 146, 148, 175
diffuse, 173, 176, 181
transmission in foliage, 174
Radiation climate, control of, 173-191
Radiation profile, 174, 183
Radiation sensors, 176
Random walk problem, 2-3
Reflectance
of soil, 187, 188, 189, 190
transmittance and canopies, 176, 177,
181, 182, 183, 187
Reflection coefficient, 185
Reflectivity of leaves, 174
Research
basic and applied, 2
fragmentation of, 5
selection of project, 1
in soil physics and technology,
aims and directions, 1-9
Respiration, 26, 61
Rhizotron, 69, 70, 72
"Roaded catchments," 19
Root(s), *see also* specific properties
branch, 58, 62
uptake of soil moisture, 89
Root concentration, 72
Root development in relation to soil
physical conditions, 47,
57-75, 68, 69
Root elongation, 58, 59, 61, 63, 64,
66-68, 73, 74, 194, *see also*
Root extension, Root development
Root environment, 26, 74, 205
Root extension and nutrient movement,
201-201, 215
Root growth, 26, 57-59, 67,
73-74, 197-199, 217

Root interception, 202
Root physiology, 26
Root respiration, 211
Root-soil environment, 57, 74
Root system, 58, 197
Root zone, 82, 101, 103
warming of, 182
Rooting depth, 157
Rooting volume, 73
Runoff, 11, 80, 81, 83, 163, 230, 231
Runoff inducement, 45

S

Saline soil
definition, 119
reclamation of, 101, 119, 120
Saline water, 14
Salinity of soil, 11, 13, 15-18, 20,
101-132
Salinity control
during crop growing period, 120-124
economic approach, 126-128
Salt accumulation, 48
Salt balance, 16, 124
Salt concentration and gradients,
109-110, 119, 148
Salt content, 44, 53, 54
Salt distribution in soil, 112-115, 117,
118, 123
Salt dynamics, 101
Salt movement (transport), 35, 40, 125
Salt uptake by plants, 115
Sand soils
advantages and disadvantages, 35
barriered, 38
improving water properties of, 35-41
Saturation, 113, 117, 120
Saturation extract, 102, 119
Seedlings, 26, 62, 66, 68, 73, 104, 175,
197, 198, 207, 208, 209, 211, 213
emergence of, 29, 30, 66, 190
Seepage, 13, 229, *see also* Percolation,
Drainage

Semi-arid regions (climate), 101
Simultaneous salt and water flow, 113
Sodic (alkali) soils, 101, 111
Sodium
  adsorption ration (S.A.R.), 111
  exchangeable, 17
  percentage, 111
Soil, *see also* specific types, properties
  fluctuations of conditions and growth
    parameters, 62, 92, 102, 103,
    112, 210, 223-232
  management of, 79, 81
Soil aeration, *see* Aeration
Soil compaction, *see* Compaction of soil
Soil conditioners, 45-48, 50-52, 216
  polymer emulsions, 50-53
  polymer solutions, 48-50
Soil conditioning, 46, 50, 52, 120
Soil crumbs, 52, *see also* Aggregates
Soil cultivation, 45, *see also* Tillage
Soil environment, 45, 53, 58, 84, 173,
  194, 216
Soil fertility, *see* Fertility of soil
Soil heat flux, 157
Soil heating, 31
Soil layering, 25
Soil management, 45, 79, 81
Soil mechanical properties, 44
Soil moisture, *see also* Soil water
  characteristic curve, 87
  control of, 92
  management of, 205-211
  redistribution of, 107, 112-114
  storage of, 82, 83, 163
  suction, 36
Soil moisture blocks, 135
Soil paste, saturated, 103, 120
Soil permeability, 17, 35, 52
Soil physical conditions, modification
  of, 45-48
Soil physical growth factors, 44-45
Soil physics, 1, 5-7
Soil-plant-atmosphere continuum, 93
Soil-plant relations, 194
Soil profile, 194

Soil reflectance, 187-189, 190
Soil salinity
  control of, 101-132
  effect on plants, 102-104, 120
    on water flow, 109-112
Soil salinity index, 121, 123, 126
Soil sampling, 135, 139
Soil solution, 54, 73, 75, 102, 103,
  105, 110, 111, 119, 123, 193,
  194, 200, 201, 215, 217
Soil stabilization, 46, 47, 49
Soil strength, 61, 66, 67-73, 74,
  see also Penetrometer resistance
Soil structure, 8, 46, 50, 53, 198
  improvement of, 43-54
  management of, 215, 216
Soil structurization, 46, 47
Soil temperature, 24-27, 44, 46, 53,
  59-61, 73-75, 188, 196, 198,
  213-215
  agronomic significance of, 25-27
  crop growth and, 23-33, 59-61
  optimization of, 23
Soil texture, 87
Soil water, 23, 44, 45, 65-67
  availability of, 44, 47, 69, 70, 92-93,
    147, 154, 155
  conservation of, 167
  control of, 18-20
  depletion of, 136, 141, 147, 148,
    157, 227
  diffusivity of, 111
  dynamics of, 24, 25
  extraction of, 89, 90
  potential, 103, 204, 205, 210
  status of, 65-67
  storage of, 11, 141
Soil-water-salt-plant system, 101
Soil whitening, 187, *see also*
    Coatings of soil and canopies
Solar radiation, *see* Radiation, solar
Solar spectrum, 179
Solute movement (flow, transport) in
  soils, 104-109, 200, *see also*
  Salt movement

Solution culture, 216
Specific surface, 194
Specific volume of soil, 54
Statistical analysis of experiments, 95
Steady-state vs. transient-state
    processes, 107, 108, 119, 125
Stomata, 94
Stomatal resistance, 146
Subirrigation, 15, 93
Suction, 36, 51, 86, 93, 94, 105, 110,
    112, 113, 123, 124, 128, 167,
    169, 207-209, 212
Sunflecks, 179
Surface charge density, 194
Suspensions, 52
Swelling of soil, 48, 110, 111

T

Temperature effects, 26
Temperature sensitivity of plants, 28
Tensiometers, 86-88, 135, 136
Tension, 38, 40, 88, see also Suction
Texture of soil, 87
Thermal properties, 24
Threshold gradient, 230
Tillage, 8, 45, 52, 53, 62, 134, 199,
    215, 216
Tillage pans, 62
Tortuosity, 105, 201
Translocation, 26, 193, 203
Transmissivity of leaves, 174
Transpiration, 14, 66, 80, 81, 83, 94,
    107, 115, 165, 167,
    168, 169, 199, 200, 204, 206
Transpiration ratio, 174
Transport law, 202
Turgor pressure, 58

U

Unsaturated soil, flow in, 109
Uptake of water and nutrients, 74,
    115, 214

USDA Irrigation Scheduling Program,
    141-158

V

Van der Waals attraction, 50
Veihmeyer concept, 93

W

Water
    conservation of, 18, 216
    desalinized, 17
    low tension, 36, 38, 40
    perched, 36, 62
    potential of, 66, 67, 103, 204,
        205, 210
    spreading of, 19
    storage of, 11, 40, 80
    uptake by plants, 115, see also
        Uptake of water and nutrients
    utilization of by crops, 91, 163-171
    viscosity of, 61, 105, 111, 201, 206
Water application efficiency, 13
Water availability, see Soil water
    availability
Water balance, 16, 19, 20, 79-100, 188
    evaluation of, 81-90
Water content, 24, 81
    distribution, 116
Water conveyance efficiency, 13
Water extraction, 117
Water holding capacity, 38
Water management, 11-22, 79, 81, 85,
    95, 139, 205, 210
Water pollution, 11
Water potential, 66, 67, 69, 195, 204,
    205, 210
Water properties of sand soil,
    improvement of, 35-40
Water requirement experiments, 118
Water retention, 35, 39, 47, 102
Water stress, 95

Water table, 15, 20, 62, 80, 125, 155, 157
Water use efficiency, 52, 79-100, 133, 163, 169
Water vapor transport, 174
Waterlogged soil, 211, 226
Watershed, 83

Weed control, 205
  permanent, 93
Wilting point, 92, 93

X

Xylem suction, 204